When a Gene Makes You Smell Like a Fish

When a Gene Makes You Smell Like a Fish

. . . and Other Tales about the Genes in Your Body

Lisa Seachrist Chiu

Illustrations by
Judith A. Seachrist

OXFORD
UNIVERSITY PRESS

OXFORD
UNIVERSITY PRESS

Oxford University Press, Inc., publishes works that
further Oxford University's objective of excellence
in research, scholarship, and education.

Oxford New York
Auckland Cape Town Dar es Salaam Hong Kong Karachi
Kuala Lumpur Madrid Melbourne Mexico City Nairobi
New Delhi Shanghai Taipei Toronto

With offices in
Argentina Austria Brazil Chile Czech Republic France Greece
Guatemala Hungary Italy Japan Poland Portugal Singapore
South Korea Switzerland Thailand Turkey Ukraine Vietnam

First published by Oxford University Press, Inc., 2006
198 Madison Avenue, New York, NY 10016
www.oup.com

First issued as an Oxford University Press paperback, 2007

Library of Congress Cataloging-in-Publication Data
Chiu, Lisa Seachrist.
When a gene makes you smell like a fish
and other tales about the genes in your body
Lisa Seachrist Chiu;
Illustrations by Judith A. Seachrist.
p. cm. Includes bibliographical references and index.
ISBN-13: 978-0-19-516994-2

1. Human genetics—Popular works.
2. Human genetics—Anecdotes.
I. Seachrist, Judith A.
II. Title.
QH431.C474 2006
599.93'5—dc22 2005031803

ISBN 978-0-19-532706-9

3 5 7 9 8 6 4 2
Printed in the United States of America

For Dan and Anya

Contents

chapter Four

chapter Five

chapter Six

Acknowledgments

This project would have been impossible without the generous help and inspiration of a host of friends, family, colleagues, and professionals. My gratitude to those who encouraged, cajoled, aided, and abetted me far outweighs the simple mention that I make of it here.

I owe an eternal debt of gratitude to my agent Jeanne Hanson for getting this project off the ground and to my friend and colleague John Travis for having the confidence in my abilities to realize this book. My partner-in-crime and friend Karen Fox kvetched over coffee, read manuscripts when needed, and spurred me to "write like a writer." Good information is the lifeblood of any undertaking of this size. Wendy Chiu, Harry Wong, and numerous scientist friends shared their knowledge, offered encouragement, and patiently granted me access to their extensive libraries whenever I needed them. Alice Chiu and Kristen Truitt provided critical advice, equipment, and the know-how that I needed to create suitable graphics for this book. I also must thank Andrea Doughty for taking my panicked phone calls and providing last minute translations. Judith Hall provided a scientific review of the manuscript. I want to thank my editor Peter Prescott for his expertise, cheerleading, gentle pressure, and basically just putting up with me during this process.

My brother Mike Seachrist, my grandmother Mary Perrin, and my in-laws Ray and Jane Chiu gave me their encouragement and support throughout this process. My father, Michael Seachrist, offered a calm confidence in my abilities that proved a priceless gift. My husband, Daniel Chiu, provided loving support and confidence every step of the way. If I can be half as supportive to him as he has been to me, I will have achieved a great good. Finally, I must thank my illustrator and mother, Judith Seachrist, without whom this book would never have been completed. Aside from the beautiful watercolors she painted for the illustrations contained in the following chapters, she gave me months of her life just to make sure this book could be born. It's a gift and sacrifice that I will never be able to repay. My only hope is that one day I can pay this generous gift forward to my own daughter, Anya.

When a Gene Makes You
Smell Like a Fish

Introduction

I was by all accounts a strange child, but for as long as I can remember, I've been fascinated with inheritance. As a five-year-old, I badgered my mother to explain to me why the twins down the street looked exactly alike, but the set two blocks over looked so different I couldn't believe they were even sisters, much less twin sisters. Shortly thereafter, my poor mother received yet another demand for explanations as I peppered her with questions asking why my brother had blue eyes and I had brown eyes. At the time, I figured that I had brown eyes like my mom because I was a girl. Therefore, my brother—a boy obviously—sported my father's striking cobalt eyes. That explanation made perfect sense to my elementary-school mind. I give my mother, a theater major in college, enormous credit because she actually tried to explain to me that my formulation was incorrect and that "genes" somehow determined whether I had blue eyes or brown eyes. She even tried to describe that blue eyes were something called recessive, and two blue-eyed parents could produce only blue-eyed children. That proved a doomed effort in part because the concept is beyond the average six-year-old.

My fascination with genetics blossomed, and one day in the third grade I found myself an eager participant in Emily Fontana's tongue-rolling experiment during recess. Emily said that if I could roll my tongue, then one of my parents could too and that I had inherited the ability from whichever parent that was. I knew I could roll my tongue, so I fidgeted all afternoon agonizing as the clock crept closer and closer to three o'clock. As soon as the bell rang, I raced out of the classroom, cutting through all the verboten backyards just to shave five minutes off the walk home. I made it home in record time only to discover that *both* my parents could roll their tongues.

Robbed of my chance to pinpoint the source of at least one of the traits I possessed, I figured it would be only a short matter of time until I had another chance. With my child's view of the world, I believed our genes told us where we came from, who we were, and what we would become. At age eight I was a genetic Calvinist clinging to the predestination written into our genes. Such black-and-white thinking is characteristic of childhood, after all. I'm not sure at what point in high school or college I began to understand and appreciate the nuance in our biological systems. I do know that by the time I first witnessed bacteria adapt to the environment (a food source lacking a critical nutrient), I was captivated by what our genes could and could not do.

Critical to my appreciation of genetics was the understanding that by and large genes don't actually *do* anything at all. When Watson and Crick described the DNA double helix that served as the body's "heritable" material, they shocked the scientific community by describing an intertwined molecule that had no known function rather than a molecule such as a protein that could actually do things. Indeed, the double helix is just two long strands of interlaced DNA that serve as the blueprint that the cellular machinery uses to make the action molecules in the cells of our bodies. If this bit of information came as a surprise, I invite you to take a look at the Genetics Primer in the back of the book. It's never a bad idea to brush off the cobwebs clinging to the information gleaned from your last biology class. However, if you can't bear the thought of being schooled, I promise you that most of the stories contained in this book will still be well within your grasp.

By the time the Human Genome Project was launched in 1990, the stories of many genes were known. But the effort to sequence the entire genome enriched that understanding and unveiled more than a few biological surprises. Still, as Francis Collins repeatedly stressed, the completion of the Human Genome Project in 2003 is just the beginning of our understanding of genetics and

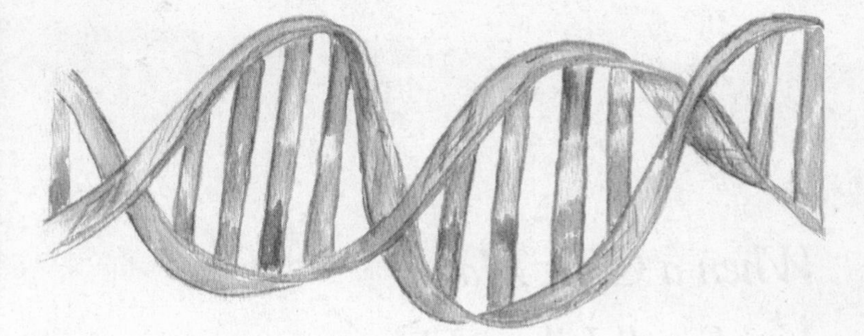

Genes are housed on the intertwined strands of DNA that make up the double helix.

biology, and indeed he was correct. Having the sequence of the entire genome hasn't suddenly cured diseases or discovered the gene for road rage. It has, however, provided the scientific community with an invaluable tool to explore the ways our genes interplay with each other and the environment.

So, read these stories not as examples of immutable truth and predestination, but as the present understanding of these genes. Who knows what future studies will unearth? The excitement comes with furthering our understanding and pinning down each new detail only to discover a new wrinkle in how our bodies use the intricate information stored in our genes.

With the human genome laid bare, scientists are rapidly discovering how the text of the genome is read in concert with things like diet, health behaviors, experiences, and other genes. With that knowledge, we gain better understanding about our similarities and differences as human beings. The short tales in this book are simply entertaining and thought-provoking snippets of the scientific discoveries driving the "Age of Biology."

Just to set the record straight, my mother had no chance of explaining to me how blue eyes are recessive to brown eyes. Eye color is governed by several genes, and is, therefore, neither a recessive nor a dominant trait. Oh, and I did pinpoint which parent endowed me with my tongue-rolling ability. Neither one. Rolling your tongue is an acquired skill not a genetic one.

When a Gene Makes You Smell Like a Fish

New Yorker Sandy Gordon knew something was amiss. It started with whispers and comments at work. Co-workers began stopping near her desk loudly asking, "What's that smell?" or "Who needs a shower?" And, she thought, yeah what *is* that smell?

And then there was that foul odor in her apartment that so annoyed her she had the building superintendent in to fix her bathroom only to find it still stank after the repairs were complete. When her nieces and nephews told her, "Aunt Sandy, you didn't brush today. Your breath stinks," she thought, perhaps, it was something she had eaten.

The awful truth came when a friend and co-worker suggested, no insisted, that they have lunch together one day. Instead of a quick bite to catch up on things, her friend told her that *she* was the source of the odor so foul and fishy it was disrupting the workplace.

"I just started crying in the restaurant," she says. "Here was the horrific realization that all those comments were about me. The smell came from me."

Devastated, Gordon went straight home and informed her company that she would be taking a three-month leave of absence to deal with her problem. "I am so grateful to my friend for dealing with it," Gordon says, "She said all the right things. She told me, I know you are a clean person; I've been to your house. It is extremely difficult to be truthful and she was."

In her early thirties, Gordon knew she hadn't been feeling particularly well. So she figured she would work with her doctors and beat whatever was making her smell. Eighteen doctors later and $28,000 out of pocket, Gordon discovered her odor condition wasn't a simple case of taking some medicine and being done with it. She learned she had a nonfatal but incurable genetic disorder: trimethylaminuria (TMAU) also known as fish odor syndrome.

As a result of genetic mutations, Gordon's body fails to produce an important liver enzyme that breaks down a smelly substance called trimethylamine (TMA), an ordinary by-product of digesting foods high in protein. For Gordon and other sufferers, this enzyme simply doesn't work and, depending on what they eat, the odor emanating from them can be barely perceptible to slightly garbagelike to the overwhelming stench of rotting fish.

Body odor isn't socially acceptable, so it's understandable that many people suffering from TMAU find themselves depressed, un- or-underemployed, isolated, and lonely. Gordon got a firsthand glimpse of the isolation when colleagues started making their "What stinks?" comments. She even had a colleague who complained to her supervisor about the odor. To the supervisor's credit, the complaining colleague was informed that only when said colleague was as productive as Gordon would the employer tolerate complaints about Gordon. Surprisingly, Gordon considers herself lucky. "I've always been a strong person and had a family that taught me to know who I was and to stand up for myself."

The first clinical case of TMAU was described in 1970 in the medical journal *The Lancet*, but literary references go back more than a thousand years. Shakespeare's *Tempest* describes the outcast Caliban, "He smells like a fish; a very ancient and fish-like smell . . . " Hindu folklore mentions in the epic *Mahabharata* (compiled around 400 AD) a maiden who "grew to be comely and fair, but a fishy odor ever clung to her."

The biochemical cause for the stink is the huge amount of TMA people with the disorder secrete into their urine, breath, and sweat. People suffering from TMAU literally walk around in a cloud of odor—at body temperature, TMA is a gas. Unfortunately, TMA is a part of everyone's life for it is naturally produced when beneficial bacteria in the human gut breakdown foods high in choline—an essential building block of proteins and a vital component for normal fat and carbohydrate metabolism as well as nerve and brain development. Foods like egg yolks, liver, organ meats, and cruciferous vegetables like broccoli and cauliflower as well as the emulsifier lecithin all contain large amounts of choline, which is broken down into TMA. The difference for people with TMAU is that they lack the enzyme needed to turn the stinky TMA into a nonodorous molecule.

Scientific and medical literature documents about a hundred or so cases of TMAU, but the real prevalence of the disease is hard to know. Researcher's estimates range from 0.1 percent to 1.0 percent of the population worldwide—in the United States this would mean as many as 250,000 individuals. John Cashman, director of the Human BioMolecular Research Institute in San Diego, notes many physicians assume odor is simply a bad case of halitosis or that the patient may be overly sensitive to odors or may be suffering from obsessive-compulsive disorder. As a result, people suffering with TMAU can wait for years to get a diagnosis.

Since the disorder was first described in the 1970s, physicians have diagnosed patients with fish odor syndrome on the basis of measuring excess TMA in their urine after they have eaten a large amount of choline—a test called a choline challenge. In 1997, physicians got another tool for confirming the diagnosis of TMAU: the gene associated with the disorder was identified: flavin-containing monooxygenase (form 3) or FMO3. This gene—housed on chromosome 1—is expressed in the liver and is part of a family of genes encoding proteins responsible for metabolizing and/or detoxifying drugs and molecules that contain nitrogen, phosphorous, selenium, and sulfur. Cashman and colleagues at the Seattle Biomedical Research Institute discovered specific changes in the FMO3 gene that could interrupt the function of the enzyme. When FMO3 can't function, people have no way to turn TMA, which contains a nitrogen atom, into the odorless substance called TMA-N-oxide.

Discovering FMO3's role in fish odor syndrome allows doctors to identify patients precisely and makes it possible to understand the nuances of FMO3 activity. The genetic discovery generated so much interest that in April of 1999, the National Institutes of Health (NIH) in Bethesda, Maryland, sponsored the first-ever symposium on fish odor syndrome. Researchers worldwide gathered to share their insights into the disease, and patients, including Sandy Gordon, were on hand to provide perspective to the proceedings.

TMAU strikes when mutations arise in the gene encoding the FMO3 enzyme. Because chromosomes come in pairs—one inherited from mom and the other inherited from dad—everyone carries two copies of the FMO3 gene. In order to suffer TMAU, a person, like Sandy Gordon, must inherit mutations in both copies of the gene. Fish odor syndrome affects the body's ability to chemically transform TMA and as such is called an inborn error of metabolism. Because it takes two mutated copies of the gene to develop TMAU, the condition is referred to as recessive. If a person has a defect in only one FMO3 gene, presumably the body's protein making machinery can use the normal gene to encode enough FMO3 to transform (or metabolize) the odor-causing TMA into its nonodorous chemical cousin.

So far, researchers have identified almost twenty mutations in the FMO3 gene that affect the body's ability to metabolize TMA into TMA-N-oxide. The gene, it turns out, is quite large. It spans twenty-seven kilobases and is comprised of nine separate sections called exons that must be correctly strung together in order for the body to produce the FMO3 enzyme. To make understanding the gene even more complicated, the gene's first section doesn't encode any part of the FMO3 enzyme and, John Cashman notes, probably serves some role in regulating whether the cellular machinery tries to make FMO3 enzyme.

However, unlike other diseases involving the obliteration of gene activity where profound effects can be seen at or shortly after birth, TMAU often doesn't develop until later in life. Sandy Gordon, for example, never had a problem with odor until she hit her early thirties. Some patients with TMAU develop symptoms when they enter puberty, while children with fish odor syndrome sometimes outgrow their odor. The disorder affects females more often than males; some of whom report that their symptoms are greater just before they menstruate.

Armed with the ability to identify both the gene and different mutations in that gene, researchers are beginning to tease out how FMO3 works in the body. Gordon's case is fairly straightforward. Genetic testing indicates she doesn't have the ability to make active FMO3. In other words, she has mutations in both copies of the gene. However, doctors and scientists still can't answer why her symptoms didn't begin until she was middle-aged.

Scientists can, however, offer explanations for why some infants develop a fishy odor and subsequently recover from the condition: it's a developmental disconnect. Before a baby is born, the fetus relies on FMO1, a related enzyme, to metabolize molecules taken care of by FMO3. The infant doesn't start making FMO3 until after it is born when it stops making FMO1. After birth, FMO1 production begins to dwindle, and FMO3 must be produced or the baby will develop a fishy odor because breast milk and formula are chock full of choline, which aids in nerve and brain development. However, the fishy odor will simply go away once the infant's body starts to produce adequate amounts of FMO3.

Researchers are also beginning to look at situations in which the amount of choline ingested simply swamps the ability of FMO3 enzyme to metabolize it. For example, patients with Huntington's disease are often given large oral doses of choline as therapy. Some have complained of a pungent odor after the treatment, likely as a result of too much choline and not enough FMO3 activity. It is also possible that people prone to overloading the capacity of their FMO3 enzyme could be carriers of mutations that cause TMAU and have only one functioning copy of FMO3. This is another area that researchers are exploring.

Cashman has been interested in FMO3 as a possible target for drug development. Most drugs are metabolized through a set of liver proteins known as cytochrome P-450 proteins. These proteins are similar in structure to hemoglobin, but rather than making sure blood is carrying enough oxygen, they are responsible for metabolizing steroid hormones and fatty acids as well as detoxifying a variety of drugs and chemical substances. Drugs that are cleared by the P-450 system tend to show considerable variability in effectiveness because the P-450 proteins are prone to drug interactions; they are activated and deactivated by small molecules. In addition, small differences in the genes—also known as genetic polymorphisms—in the P-450 proteins also dictate how well a drug works from one individual to the next.

Hormones influence FMO3 activity as evidenced by sufferers who experience odor associated with their menstrual cycles. Even so, FMO3 activity is apparently not induced by other chemicals or drugs. Cashman points out that all variability seen in FMO3 activity can be viewed as the result of genetic variation. That fact opens a new and potentially valuable avenue for drug development.

Cashman sees drugs metabolized by FMO3 as inherently more predictable because they will cause far fewer drug interactions. As a result, most people could take a drug targeted to the FMO3 system and enjoy fewer side effects, as well as better efficacy, than yet another drug metabolized via the P-450 system.

Most people, that is, except those who suffer from TMAU. Drugs targeted to FMO3 would likely be ineffective or even dangerous for patients who have no FMO3 activity. In a cruel twist, discovering the gene that makes people smell fishy may actually end up helping those who have normal FMO3 genes.

Still, patients like Sandy Gordon believe that with enough research funding and scientific interest in this peculiar condition, there will be a cure for fish odor syndrome. In the meantime, doctors, researchers, and patients search for ways to manage the smell. Many of them end up at the Monell Chemical Senses Center in Philadelphia to see odor specialist George Preti about ways to dampen the stench. Preti isn't a physician; he's an organic chemist who has been studying human body odors for thirty years. "I'm not here to be nice. My job is to find the cause of the odor and figure out how to reduce symptoms and take care of it."

Preti says that often by the time patients see him, they have tried every odor-control product that can be scrounged from drugstore shelves and the Internet. Preti's first order of business is to determine what exactly is causing the odor problem. "We have to test to know what we are dealing with," he notes. That means conducting a choline challenge: patients ingest an enormous amount of choline, then return the following day to see if they excrete copious amounts of TMA. If they do, Preti then refers them for genetic testing.

Preti advocates using charcoal and/or copper chlorophyllin daily as a means of absorbing the odor. The products are used in nursing homes to control odors in patients with incontinence and colostomies. A study of Japanese patients with TMAU indicated charcoal and copper chlorophyllin reduced the amount of TMA excreted in the urine to normal levels. In addition to the charcoal and copper chlorophyllin, Preti says the antibiotic metronidazole may also offer some people relief from their symptoms.

Even though patients are eager for a cure, Preti doesn't think TMAU lends itself to a simple cure. He notes that it is unlikely that a simple enzyme replacement will work as it has in diseases like Gaucher's disease and hemophilia. In those diseases, the enzyme being replaced is a digestive or blood-borne enzyme. The FMO3 enzyme is housed inside sacs called macrosomes in liver cells, which makes it very difficult to get the enzyme where it needs to go.

Yet, patients like Gordon remain hopeful that the scientific community will come up with some solution for the problem. She says that, for her, chlorophyll and avoiding certain foods is helpful, but those measures don't take care of the odor entirely and "some days the odor is worse than others."

Odor or not, Gordon hasn't taken her condition lying down. She has become an activist for people suffering from TMAU. At the first NIH conference devoted to TMAU, Gordon started the Trimethylaminuria Support Group—a move she credits as the true beginning of her healing. When the NIH hosted a second conference, Gordon's group was one of the sponsors. Gordon relishes her role as an outspoken activist for people with TMAU. "I tell people they have every right to do what they want to do. I also tell them to get psychological counseling. Nobody wants to have an odor—it just isn't socially acceptable to have an odor. But you can manage it."

Until science finds a way to eliminate that odor, that is just what Gordon and her fellow sufferers will have to do: manage.

chapter One

It Takes Two to Tango

In many ways, the story of fish odor syndrome fittingly illustrates a number of basic principles about inheritance. For example, people with two defective copies of the FMO3 gene will have the syndrome and the concomitant odor. Those who are fortunate enough to have two normal copies of the gene needn't worry about the syndrome at all. This type of inheritance pattern is an example of Mendelian inheritance because it follows the observations detailed by the Austrian monk Gregor Johann Mendel.

Any discussion of modern genetics and inheritance must start with the work of Mendel and his cultivation of garden peas. Mendel was born in Heinzendorf, Austria, on July 22, 1822, the son of peasant farmers. A gifted student, he began studying at the St. Thomas Monastery of the Augustinian Order in Brunn (now Brno, Czech Republic) in 1843 when his parents could no longer afford to pay for his education. In 1847, he was ordained.

It was clear that the young Mendel was better suited to intellectual than pastoral pursuits, and he was assigned to teaching duties and sent to the University of Vienna to become a math and biology teacher. Fortunately for the future field of genetic science, he failed the teaching certification exam repeatedly and was forced to return to the monastery in Brunn to teach part time where he began cultivating and studying his peas.

Mendel chose to focus on seven observable characteristics of the pea plants: round or wrinkled seeds; yellow or green seed interiors; purple or white flowers; yellow or green pods; inflated or pinched pods; terminal or axial flowers; and long or short stems. By selectively breeding for these characteristics, he was able to produce peas that continuously "bred true" for each characteristic. For example, white-flowered plants always produced white flowers, and purple-flowered plants always produced purple flowers.

Things got interesting when he bred these pure characteristic lines to each other, i.e., purple-flowered plants to white-flowered plants. In the first generation, all of the plants produced purple-flowers; not a paler shade of purple like lavender, but a purple the same shade as the purple plants that served as its ancestors. Mendel interpreted this result to mean the purple flower color was dominant to the white flower color. At this point, Mendel took advantage of the reproductive capacity of garden peas—these plants can reproduce by fertilization between two pea plants or by self-fertilization. He allowed all of the resulting purple-flowered plants to self-fertilize. Most of the resulting pea plants produced purple flowers, but some of them produced white flowers. Somehow, the white character had reappeared.

By counting the number of purple flowers vs. white flowers, Mendel discovered that for every three purple-flowered plants, he found one white-flowered plant. What's more, this 3:1 ratio was consistent for all seven of the characteristics he was studying.

Further experimentation showed that not all purple flowers were alike. In fact, he found one-quarter of the plants produced pure-breeding purple flowers, one-quarter produced pure-breeding white flowers and one-half of the flowers were "impure" purple flowers. These ratios held for all seven characteristics he studied. In addition, Mendel took his experiments even farther by studying the inheritance of two characteristics, and he found that the characteristics were inherited independently. Mendel

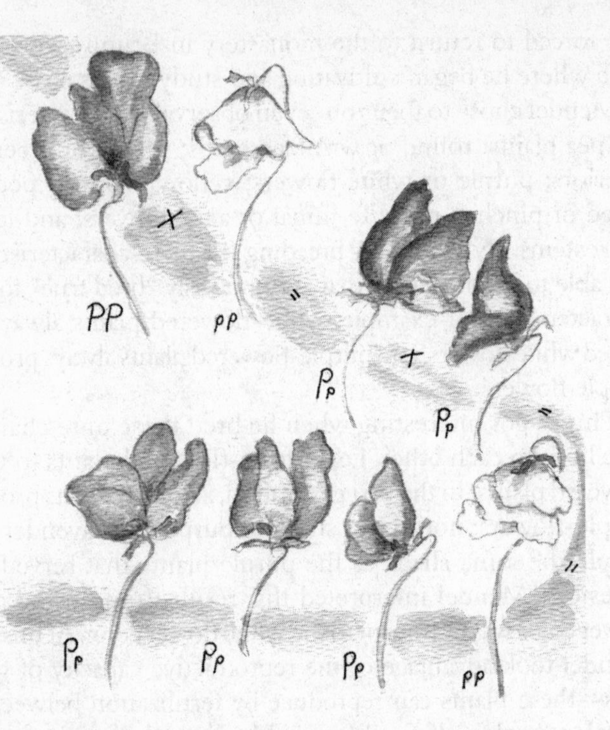

Mendel's experiments with peas established the rules for inheritance. Working with purple-flowered and white-flowered sweet peas, Mendel demonstrated that purple-flowered plants were dominant to white-flowered plants. When he crossed purple-flowered plants (PP) with white-flowered plants (pp), the first generation (which received one gene from each parent) produced only purple-flowered plants (Pp). Crossing the offspring of the white- and purple-flowered plants produced three purple-flowered plants (one PP, two Pp) and one white-flowered plant (pp).

presented his papers in 1865 at two meetings of the Natural History Society of Brunn; his paper titled "Experiments on Plant Hybridization" was published in the *Proceedings of the Natural History Society of Brunn* in 1866.

Before scientists knew that DNA was the source of genetic material, before the word gene had even entered into the scientific vernacular, Mendel precisely described the inheritance of discrete characteristics in a way that holds true for many genes today. Nevertheless, his discoveries were destined to be overlooked for more than thirty years. In fact, Mendel died in 1884 without ever receiving the scientific recognition he deserved.

Mendel's work was rediscovered in 1900 independently by Carl Correns in Germany, Hugo de Vries in the Netherlands, and Erich von Tschermak-Seysenegg in Austria. Each of these men recognized that Mendel had described how specific characteristics were passed from parent to offspring.

However well Mendel described the discrete inheritance patterns of some sweet pea traits, the situation often isn't so tidy. For example, sometimes an intermediate situation akin to the production of a lavender-flowered plant appears. Fish odor syndrome highlights this occurrence. When a person carries one normal copy and one defective copy of the FMO3 gene, they can experience no odor, occasional odor, or odor as bad as those who carry two defective copies of the gene. These people are often called carriers, but scientists refer to them as heterozygotes.

Whether an FMO3 gene heterozygote experiences symptoms of fish odor syndrome depends upon a number of different factors, such as how much active enzyme the person makes, whether he or she eats a diet very high in choline, does this person take certain drugs? In other words, a heterozygote's *environment* can determine whether or not he or she experiences odor. That interplay between a person's genetic makeup (his or her genotype) and the environment defines much of the way in which genes are expressed as traits (phenotypes) such as, in the case of fish odor syndrome, odor.

Most common human diseases like heart disease and arthritis include genetic components influencing the outcomes of these conditions; that's the reason these types of diseases tend to run in families. Because common diseases arise from many intricate genetic and environmental interactions, discovering the genes important in these diseases is a complicated process that has only recently begun to bear fruit.

A century ago, when scientists first began identifying human genetic diseases, the tools for teasing out tangles of genetic influences such as recombinant genetics, the complete human genome and the understanding of DNA as the home of heritable traits had yet to be discovered. Physicians of the day identified genetic diseases with the tools of observational medicine and the newly burgeoning field of biochemistry. Because biochemical analysis was first applied to urine and not to blood, most clinical chemistry undertaken at the time involved examining urine to look for unusual substances or excesses of normal substances. In some ways these early geneticists were the ultimate piss-prophets: the first genetic disease ever identified was the one that caused people to have black urine, and urine analysis identified the genetic problems for many mentally retarded individuals.

Irrespective of their seeming fascination with urine, these early genetic pioneers identified very rare genetic diseases that displayed distinct and sometimes terrible phenotypes. These were the genetic conditions that most resembled the rules of inheritance described by Mendel with his peas: autosomal (meaning not sex-linked) recessive and dominant disorders. Genetic disorders closely following Mendel's rules of inheritance are pretty much the low-hanging fruit on the human genome because the effects tend to be dramatic and the inheritance pattern is clear. Still, the stories of these genes illuminate core genetic principles that still ring true today.

To start at the beginning, we need to go back more than one hundred years to examine the first condition recognized as genetic in nature—the recessive disease known as the black urine disease or alkaptonuria.

The First Gene

In 1584, a boy who passed inky urine came to the attention of Wilhelm Adolph Scribonius, a Marburg physician who as well had the dubious distinction of being a leading proponent of the water test for witches. The boy was otherwise healthy, and his condition proved nothing more than a medical curiosity. Over the centuries, this unusual trait was noted by other physicians and scientists. In 1859 Carl H. D. Boedeker discovered certain chemical compounds such as sodium hydroxide could trigger some people's urine to turn black and dubbed the condition alkaptonuria. In 1891, M. Wolkow and E. Baumann unearthed the cause of that blackening: excess alkapton (now called homogentisic acid) turning dark brown or black when it comes into contact with air.

Alkaptonuria, while biochemically interesting, wasn't particularly important medically when the English physician Archibald Garrod became interested in the biochemistry of the disorder in the late 1800s. Passing urine that turned black may be a little odd, but the patients didn't appear to be harmed. And, aside from the shock new parents got when they changed an ink-stained diaper, the condition wasn't considered a cause for alarm.

Nevertheless, Garrod chose to study these patients and in the process applied the newly rediscovered Mendelian laws of inheritance to human diseases and conditions. Rather than simply studying individuals with the disorder, Garrod studied their families as well. Because alkaptonuria is usually diagnosed when an infant's diaper turns black or dark brown, Garrod could be certain that the condition was present from birth and was not some sort of condition that could be picked up later in life. In addition, he noted that where there was one alkaptonuric in a family, he usually found others. The condition was, in fact, inherited.

The rediscovery of Mendel's work proved critical to Garrod's understanding of alkaptonuria and metabolic processes in man. As Garrod collected information about patients with alkaptonuria, he noted a couple of curious things. The first was that quite often the condition could be found in two or more brothers and sisters whose parents were normal, and the condition wasn't known in any of their forefathers. In other words, the black urine occurred spontaneously and usually affected more than one child in the family. In addition, many people who had alkaptonuria were the children of first-cousin marriages—one study showed that out of four British families with alkaptonuric children, three were first cousin or consanguineous marriages.

Garrod remarked on the high level of consanguinity in his ground-breaking 1902 paper in the British medical journal *The Lancet*. However, he also noted "the proportion of alkaptonuric families and individuals who are the offspring of first cousins is remarkably high . . . it is equally clear that only a minute proportion of the children of such unions are alkaptonuric."

Mere consanguinity couldn't be the cause of the black urine. Knowing that the condition was in fact exceedingly rare, Garrod posited the answer most likely was "some peculiarity of the parents, which may remain latent for generations, but which has the best chance of asserting itself in the offspring of the union of two members of a family in which it was transmitted." And this is where Mendel's laws of heredity proved useful.

Garrod's colleague William Bateson, the first English-speaking scientist to recognize the importance of Mendel's work, suggested to him that Mendel's peas may provide a clue as to what was happening in these families. Mendel studied discrete characteristics that offered mutually exclusive phenotypes. Garrod considered the presence of black urine disease and the absence of black urine disease as two mutually exclusive phenotypes: having large amounts of homogentisic acid in the urine was the recessive characteristic to the "normal" condition with minute amounts.

Bateson had restated Mendel's laws to note that when an egg or a sperm (a gamete) bearing a recessive characteristic combined with one of the dominant type, the person born of that fertilized egg, or zygote, would produce the dominant characteristic. And when a recessive gamete met with another recessive gamete, that person would produce the recessive characteristic. In other words, it took the melding of an egg harboring the recessive characteristic with a sperm harboring the same recessive characteristic in order for that characteristic—in this case alkaptonuria—to be seen in the offspring. If the recessive characteristic was very rare, the recessive condition would be extremely uncommon. Bateson pointed out that first cousins were most likely to share similar gametes—even if they are rare in the population as a whole. As a result, first-cousin matings were the most likely to enable the fertilization between eggs and sperm carrying the recessive characteristic resulting in the appearance of such rare conditions as alkaptonuria.

Garrod recognized that Mendelian inheritance explained why normal parents sometimes produced alkaptonuric children and why parents of alkaptonuric children were more likely to be first cousins. In addition, because only one of the five children of two alkaptonuric fathers also had alkaptonuria, it was clear the condition couldn't be directly inherited.

While Mendel's laws of inheritance provided a good explanation of HOW alkaptonuria was inherited, they did nothing to explain WHAT exactly was being inherited. Here, too, Garrod provided a critical insight. Garrod knew that alkaptonurics' characteristic dark-staining urine occurred as a result of excess homogentisic acid (HGA) in their urine. In studying the biochemistry of the disease, Garrod came to focus on how the body breaks down proteins. While DNA is comprised of individual units called nucleotides, proteins are comprised of amino acids. Garrod deduced that the excess HGA, a chemical cousin of the amino acids phenylalanine and tyrosine, was actually the breakdown product of those two amino acids.

All organisms, whether plant, animal, or bacteria, devote a tremendous amount of resources not just to breaking down and excreting excess cellular components but also to synthesizing components they lack. Combined with the production and use of energy, these are the processes of metabolism that ensure an organism's well-being. Archibald Garrod determined that in alkaptonuria a critical step in the breakdown pathway for the amino acids phenylalanine and tyrosine was either missing or disabled. He surmised the enzyme missing was responsible for turning the colored HGA into a colorless form. As a result, alkaptonurics accumulate so much HGA that the excess must be excreted by the kidneys. Garrod termed alkaptonuria an "inborn error of metabolism."

In 1908, Garrod published his first edition of *Inborn Errors of Metabolism* wherein he noted the accumulation of HGA in urine represented a defect in a single enzyme. Garrod's work provided the first suggestion of the one gene–one enzyme hypothesis.

That work, like Mendel's, was greeted coolly by the scientific community and was largely overlooked until the 1940s when George W. Beadle of the California Institute of Technology and Edward L. Tatum of the Rockefeller Institute for Medical Research independently proposed the one gene–one enzyme hypothesis. Beadle generously praised Garrod's work in a speech accepting his 1958 Nobel Prize in Physiology and Medicine. As Tatum noted, alkaptonuria was the first of many diseases known as inborn errors of metabolism, "there are now many diseases described as such; in fact, they have come to be recognized as a category of diseases of major medical importance."

It wasn't until the mid-1950s, that the actual enzyme responsible for alkaptonuria was discovered. Bert N. La Du and colleagues at the NIH in Bethesda, Maryland, found alkaptonurics lacked a liver enzyme dubbed homogentisic acid oxidase, or HAO. In 1996, Jose Fernandez-Canon and colleagues in Spain found the gene for HAO on chromosome 3. Since that discovery, more than forty different mutations in the HAO gene associated with alkaptonuria have been identified.

At the time he described alkaptonuria as an inborn error in metabolism, Garrod proposed many more such defects were likely to be found. He maintained that they were apt to be rare, inherited in a Mendelian fashion, and mostly harmless. Garrod got three out of four correct. Other inborn errors of metabolism such as galactosemia (an accumulation of the sugar called galactose) and phenylketonuria (an accumulation of the amino acid phenylalanine) were discovered, and as Garrod predicted, they proved rare and recessive. It's Garrod's assertion that inborn errors of metabolism are harmless that ultimately proved incorrect. Most diseases classified as inborn errors of metabolism are devastating. If untreated, galactosemia causes liver and brain damage; phenylketonuria results in profound mental retardation.

While the alkaptonurics Garrod studied appeared to suffer only from black urine, it's now known that not all of the excess HGA they produce is excreted by the kidneys. Some of that excess HGA accumulates in cartilage and stains it black—a condition known as ochronosis. As the joint cushioning cartilage darkens, the joints become arthritic, and after the age of thirty, most alkaptonurics begin to suffer from osteoarthritis, which causes pain, stiffness, and swelling in joints such as the spine, knees, hips, and shoulders. In some cases, alkaptonuria isn't even diagnosed until the patient seeks treatment for chronic joint pain. In addition, HGA accumulates in the cardiovascular system as well, with the heart's aortic and mitral valves being the most affected. The deposited HGA can cause the valves to harden and the deposited pigment may also cause atherosclerotic plaques. Alkaptonurics may also experience darkening of the cartilage in their ears and the whites of their eyes.

Unfortunately, there aren't many treatments for these patients. Maintaining a low protein diet, particularly one that avoids the amino acids phenylalanine and tyrosine by limiting foods such as eggs, meat, and nuts, can reduce the accumulation of HGA and as a result slow the development and progression of ochronosis and arthritis. Some studies have shown that high dose vitamin C (ascorbic acid) in combination with a low protein diet is even better at slowing the accumulation of HGA. Researchers are also looking at the drug nitisinone—used in the treatment of another inborn error of metabolism tyrosinemia type 1—as a means of reducing HGA levels, but nitisinone therapy is still being studied because it causes the accumulation of tyrosine, which can lead to neurological problems.

Even though alkaptonuria can cause significant problems later in life, it is, in relative terms, a benign "error in metabolism." The more than 350 other inborn errors of metabolism tend to cause severe problems such as mental retardation, deafness, and blindness as well as kidney, liver, and cardiovascular problems. While Garrod failed to understand the medical importance of his newly identified inborn errors of metabolism, he did successfully identify a major area of human genetic variation.

Phenylketonuria and the First Genetic Test

The mother in Asbjørn Følling's waiting room simply wouldn't give up. She had two mentally retarded children and wanted to know why both her son and daughter suffered from the same feeblemindedness. Why did a peculiar smell cling to them? Could Dr. Følling help her son and daughter when so many others had failed?

Følling, in 1934, was a Norwegian doctor who'd trained first as a chemist. He agreed to examine the woman's two children who were, at the time, four and seven. Upon first observation, he saw nothing unusual in the children besides their clear cognitive impairment. But, as his son Invar Følling accounts in a special edition of *Acta Pediatrica*, Følling needed to complete additional biochemical tests.

The first was a routine urine test used to detect ketones—chemicals the body makes when the body begins to break down proteins. That test simply requires that a little bit of the chemical ferric chloride be added to urine. In normal individuals the ferric chloride would stay brown. When diabetics fail to keep their sugar levels controlled, they excrete ketones in their urine because their bodies have resorted to breaking down their own muscles. Among ketone-excreting patients, ferric chloride turns their urine a burgundy-purple color. When Følling added ferric chloride to the urine from these two children, it turned dark green. Whatever these children were excreting in their urine, Følling had never seen it before.

Using the skills in organic chemistry that he had honed before becoming a physician, Følling extracted the molecule and discovered it was a chemical known as phenylpyruvic acid. Armed with a measurable trait, Følling searched throughout Norway to find other children who excreted phenylpyruvic acid in their urine and suffered the same profound mental retardation as those first two children. He found thirty-four cases in twenty-two families and showed the condition was an autosomal recessive disease—a child had to inherit a defective gene from both parents. In 1938, Følling with a colleague realized the phenylpyruvic acid came from the kidneys excreting an excess of the amino acid phenylalanine and speculated the children had a defect in their bodies' ability to metabolize the amino acid.

In other words, however the excess phenylalanine was causing the mental retardation, it was yet another example of an inborn error of metabolism. Følling's assertion was born out in 1947 when American scientist George Jervis identified the problem as a defect in phenylalanine hydroxylase—an enzyme that converts phenylalanine into the amino acid tyrosine.

Without the ability to break down phenylalanine, the amino acid builds up in the bloodstream and spills out into the urine. Elevated levels of phenylalanine in the bloodstream disrupt such vital brain development processes as coating the nerve cells with a protective myelin sheath. As a result, victim's brains fail to develop normally, and they are left severely mentally retarded. The disorder was dubbed phenylketonuria (PKU), referring to the ketone breakdown products excreted in the urine.

PKU's devastating effects on the still-developing nervous system starkly illustrates that, despite Archibald Garrod's assertion, the consequences of inborn errors of metabolism are far from uniformly benign. While it affects only about one out of 11,000 children born in the United States, without treatment, these children are destined to be severely disabled.

At the time medical science established that excess phenylalanine could cause neurological damage, however, physicians didn't have a treatment to offer to prevent the neurological decline. American biochemists George Jervis and Richard Block as well as Britain's Lionel Penrose began proposing that affected infants could be treated with a low-phenylalanine diet that would ultimately stymie their mental decline. The idea failed to take hold in the 1930s because the proponents assumed the synthetic food would be prohibitively costly.

In 1951, English biochemists Louis Woolf and David Vulliamy decided to test the dietary theory in three small children suffering from PKU. All showed some improvement. Other British and American researchers followed suit, but the gains showed by the treated children were relatively small, leading most scientists to conclude that any therapy needed to be started shortly after birth in order to prevent irreversible brain damage.

The challenge to identifying children at such an early age was that the ferric chloride test Følling had used to discover the phenylalanine in urine wasn't reliable until six to eight weeks after the birth of the child—a point when irreversible brain damage may have already occurred. Mass screening, and therefore treatment, couldn't become a routine practice until physicians had access to an improved testing method.

Enter Robert Guthrie, a microbiologist and physician practicing in Buffalo, New York. Guthrie's son and niece suffered mental retardation, and his niece's condition was the result of PKU. Guthrie developed a test for PKU involving a blood specimen on a disk of filter paper. He placed the specimen disk on a culture dish of bacteria that couldn't grow without additional phenylalanine. If the blood came from a child with PKU, the high levels of phenylalanine in the bloodstream would allow the bacteria to grow.

Guthrie's test had the advantage of being able to detect PKU by the third day of life. Presumably, infants identified so early could be placed on a low phenylalanine diet soon enough to prevent mental retardation. The test and treatment held so much promise that in 1963, the U.S. government was promoting PKU testing for all newborns. By the 1970s—two decades before the gene encoding the enzyme phenylalanine transferase was isolated by Savio Woo at Mount Sinai School of Medicine—PKU testing was routine across the country.

The Guthrie test, which involves pricking the heel of a newborn infant, was the first screen developed for a genetic disease. More than 150 million infants now have been screened, and over 10,000 have been detected with PKU and have been treated with a low phenylalanine diet.

The story of phenylketonuria appears to be a rousing success story for the use of genetic testing and medical intervention to prevent disease. It is an incomplete success at best. There is no doubt that children have been spared from serious mental defects, but it hasn't been without cost.

A positive test for phenylketonuria sets the family and infant down a long and difficult path. While the special diet is critical to prevent mental retardation, it is enormously restrictive. Because phenylalanine is present in protein, high protein foods aren't allowed. Patients avoid meat, milk, eggs, cheese, beans, peanut butter, and nuts. That means children on this diet must forgo the pleasures of normal food such as pizza and ice cream. Even simple foods like bread and pasta must be low-protein varieties. And, in order to make sure they have enough protein and other vitamins to maintain good health, they are required to drink a phenylalanine-free formula that is so offensive in taste and odor even researchers in the field acknowledge it's hard to stomach.

When the screening test for phenylketonuria first began identifying people with the condition, physicians assumed patients would need to stay on the diet only until their brains had fully developed by the age of five or six. Diane Paul, a political science professor at the University of Massachusetts at Boston, has pointed out in an appendix to the Final Report of the Task Force on Genetic Testing—a National Human Genome Research Institute project—that because widespread screening was initiated before there was proof that the diet would work, physicians didn't really know if "only infants and young children needed to maintain the restricted diet." Nevertheless, Paul notes that the optimistic assumption was often presented as fact to the public.

However, such rosy prognostications haven't borne out. Adults who stop the restrictive diet don't suffer mental retardation, but a follow-up to the United States Collaborative Study of Children Treated for Phenylketonuria (PKUCS)—a project begun in 1967 by the U.S. Children's Bureau—documents that those who went "off-diet" are more likely to have a lower intellectual capacity than those who "maintained" their diets. In addition, patients who didn't continue the phenylalanine-restricted diet were more likely to suffer from eczema, asthma, mental disorders, behavior problems, headache, and hyperactivity as well as lethargy.

Cognitive decline and emotional problems aside, adult female phenylketonurics have another consideration—children. Happily, women who once were unlikely to have children because they were severely mentally retarded are now leading full lives including getting married and raising families. There is, however, a dark side to this story. In 1980, Roger Lenke and

Harold Levy reported that mothers who don't maintain strict control over their levels of phenylalanine are more likely to bear children with mental retardation, microcephaly (small brains), and congenital heart disease. In other words, all the postnatal heel sticks and Guthrie screening will have had no benefit whatsoever if the disease is just postponed until the next generation.

What's more, their results hinted that returning to the PKU diet once a woman knew she was pregnant wasn't likely to prevent birth defects in the children. Instead, they suggested that mothers establish strict control over their condition *before* they became pregnant. Subsequent studies confirmed Lenke and Levy's concern. Women with PKU who wish to have children need to have their diet under control *before* ten weeks gestation and preferably even before they become pregnant.

Maintaining control over PKU during pregnancy can be particularly difficult. Because of surging energy and nutrient needs, a pregnant woman must increase the amount of phenylalanine-free formula she ingests. But a nauseated woman isn't going to relish consuming additional quantities of an unpalatable substance. In addition, because maternal PKU is a relatively new phenomenon, many physicians aren't experienced in helping the mother keep her phenylalanine levels under control.

Having the means to test for and treat PKU is nothing less than a stunning achievement. Even so, the current methods for treating PKU, as well as a number of other inborn errors of metabolism such as tyrosinemia I and maple syrup urine disease are far from ideal. Levy has noted that any new means that will improve protein tolerance for phenylketonurics would lift a significant burden from patients and their families. Researchers around the globe are exploring whether a bacterial enzyme that breaks down phenylalanine could be eaten with meals to lower the phenylalanine levels in foods before they are absorbed by the body.

Even though the fix is imperfect, PKU screening has rescued thousands of young people from certain disability. And, the experience with PKU screening has led to testing newborns for a number of other genetic disorders.

Maple Syrup Urine

Nestled in the heart of Lancaster County, Pennsylvania, stands the Clinic for Special Children, a gray clapboard building raised in the traditional way with horses, ropes, and strong backs by the Plain People in the area. The building fits into the simple ways of the Old Order Amish and Mennonites who've made their home in the region since the early 1700s.

The clinic itself is an interesting paradox: it's a facility that seeks to incorporate the practice of modern genetics with general medical care for communities that eschew modernity. An incongruous arrangement to be sure, but both the Amish and Mennonite communities in the area carry a significant burden for rare genetic diseases. And, for these proud, self-sufficient communities, the clinic is a means to maintain their tradition of taking care of their own. It is here where the Mennonites of southeastern Pennsylvania bring their children when the rare genetic disorder called maple syrup urine disease (MSUD) strikes.

Isolated both by geography and culture, the Amish and Mennonite communities have proven a treasure trove for studying genetic disease. The communities keep excellent records and have large families as well as having very similar diets, lifestyles, and other environmental exposures. All of these characteristics make it easier for researchers who are trying to understand genetic diseases. However, the biggest advantage comes not from shared lifestyles but from a random population phenomenon known as the "founder effect."

As much as evolutionary theory dictates that the genes we inherit are the result of selective pressure over the ages, sometimes certain genes persist in a population simply out of chance. The founder effect is largely a function of that chance. Throughout a large population, any single gene comes in a number of different flavors; to be more specific, these are called alleles. Rare genetic diseases result when a person inherits two alleles that alter the normal function of whatever the gene is ordinarily supposed to produce. In large populations, the frequency of such deleterious alleles is low enough to keep these diseases extremely

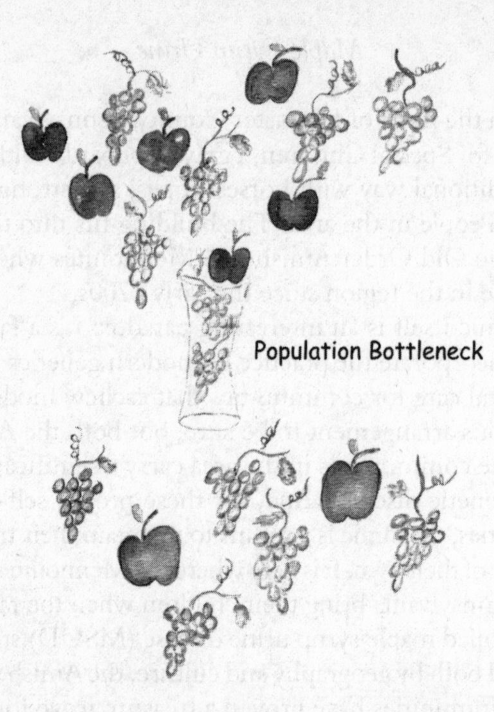

Genes aren't immutable sequences. They are often comprised of subtle differences akin to flavors. These differences are called alleles, and for any given gene those alleles appear in a population at certain frequencies. When a population is reduced in size, or goes through a "population bottleneck," those frequencies change depending on exactly which alleles make it through the bottleneck. In the illustration, the "apple" and "grape" alleles are present in a 50:50 ratio at the beginning. After going through the bottleneck, the grapes outnumber the apples almost 2:1. Scientists call this the founder effect.

rare. However, if a community becomes isolated either because of its geographic location or social customs, the frequency of certain alleles depend entirely on the genetic makeup of the founders of the group. That is, whatever alleles, beneficial or deleterious, are present among the founders will be enriched in frequency amongst their descendants—hence the founder effect.

As a result of the founder effect, communities such as the Old Order Amish and Mennonites, Ashkenazi Jews, and the populations of Finland and Iceland have a higher incidence of rare genetic diseases specific to their populations. That's not to say the founder effect only confers an increased risk of genetic disease; these populations will be enriched in *both beneficial and deleterious* alleles possessed by their founders. In genetics as in life it's sometimes much easier to see what's wrong rather than what's right.

In some ways you could say the founder effect is what prompted D. Holmes Morton and Caroline S. Morton to establish the Clinic for Special Children in the first place. The Amish and the Mennonites in the area don't intermarry and as such are reproductively separate from each other to the extent that as a result of the founder effect each group is at high risk for completely different diseases. The two researchers originally set out in 1989 to diagnose and treat the Mennonite children suffering from MSUD and the Amish children suffering from glutaric aciduria type 1 (GA1).

The clinic has since branched out to include a number of other rare genetic diseases including ataxia-telangiectasia—a multisystem degenerative disease causing loss of muscle control, immune system problems, and susceptibility to cancer—congenital deafness, and familial manic-depressive illness, all of which are also more common in these populations. But the two conditions the Mortons first intended to address offer some interesting examples of the founder effect.

Among the Amish living in southeastern Pennsylvania, glutaric aciduria type 1 (a defect in the metabolism of the amino acids lysine and tryptophan) is a relatively common occurrence; 10 percent of the Amish carry a defect in the glutaryl CoA-dehydrogenase enzyme, which leads to the disease. The condition is sometimes called Amish cerebral palsy because it causes repetitive muscle contractions that result in twisting and jerking motions. However, it also results in progressive neurodegeneration, which causes swallowing and respiratory problems. While the disease affects greater numbers of the Amish as compared to the U.S. population as whole, the specific allele that causes GA1 in the Amish is a common one that also causes the disease in Europe and the United States.

Luck took a holiday in the case of maple syrup urine disease (MSUD), however. The allele responsible for causing MSUD among the Mennonites in the area, while not unknown, is extremely rare worldwide. And because MSUD is a highly variable disease, different alleles will cause more and less severe phenotypes. The allele that took hold among the Mennonites causes one of the more serious forms of the disease. The story of how such a deleterious allele took hold in today's Mennonite population is one of repeated chance survival.

Maple syrup urine disease was first described in 1954 by John Menkes, who was at the time a medical intern at Boston Children's Hospital, and his colleagues Peter Hurst and John Craig. The intern, junior resident, and pathologist had been introduced to a mother who'd lost two children very early to a neurodegenerative disease thought to be associated with excessive bilirubin, the yellow-colored pigment found in bile that, in excess, causes jaundice. Menkes noted that the mother said the two infants who'd died had a peculiar odor, while the child who'd survived infancy did not.

Menkes's young patient eventually developed the odd odor and soon thereafter started to develop neurological symptoms: lethargy, irritability, unusual muscle-tone, seizures, difficulty breathing, and signs of brain swelling. The child eventually died.

In 1978, Menkes recounts asking everyone in the hospital at the time to identify the smell in the child's urine only to get the description that it smelled like maple syrup. Determined to identify what was causing the smell, Menkes and Hurst sniffed their way through a Harvard University organic chemistry stockroom comparing the smell of known chemicals to the child's urine. They decided the odor most closely resembled malic acid and assumed they'd identified a condition where an enzyme deficiency resulted in the accumulation of malic acid.

Menkes ultimately discovered that one of the key ingredients in artificial maple syrup was a type of chemical known as a ketone with a structure similar to the amino acids leucine, isoleucine, and valine. These amino acids are called the branched chain amino acids. Menkes noted when he came across the disease again in 1957 that he was able to identify branched chain ketoacids in the urine as the source of the smell and to determine that maple syrup urine disease is a defect in the body's ability to use the "branched chain" amino acids properly.

MSUD, a rare recessive condition, occurs when a person inherits defects from both parents in any part of a four-protein complex that breaks down these amino acids. The disease itself is highly variable. Researchers believe much of that variability comes from inheriting defects in different components of the enzyme complex.

For the most part, MSUD is exceedingly rare. The disease afflicts approximately one out of every 180,000 newborns in the United States. In Old Order Mennonites in southeastern Pennsylvania, however, the condition is much more common, affecting one out of every 200 newborns.

Classic MSUD is the most common and severe form. All newborn infants spend their first few days breaking down protein in their own bodies as a response to the sudden loss of direct to the bloodstream nutrition from mom. Because they have no way to process branched-chain amino acids, infants suffering from classic MSUD experience a rapid increase in those amino acids during that time. Within forty-eight hours, the infants will become irritable, lethargic, and difficult to feed. By four to seven days, they will start to show neurological symptoms such as breathing difficulties, problems with muscle tone, seizures, and brain swelling. Once these symptoms appear, the child is in a crisis that can quickly lead to brain damage and death.

For the Old Order Mennonites, a change in a single nucleotide for the gene that encodes one of the proteins comprising the branched-chain alpha-keto acid dehydrogenase (BCKDH) complex causes MSUD. The Mennonite version of MSUD arises when that single change in the DNA code causes the cellular machinery to stop manufacturing one of the proteins in the BCKDH complex. All Mennonite patients to date have inherited two copies of this mutant allele that results in classic MSUD.

Understanding how such a rare gene became so common in the Mennonite community of southeastern Pennsylvania requires a look at some history. The Anabaptist movement, from which both the Amish and Mennonite denominations descend, originated in 1525 in Switzerland. As the name indicates, the Anabaptists held that only adult believers could make the decision to join the church and be baptized. In addition, they were vigorous proponents of the separation between church and state. At the time, the concept of a separation between church and state was a radical idea, and the Anabaptists found themselves often in conflict with both the church *and* the state.

The group was forced from Switzerland to Germany where they were tolerated as long as they paid heavy taxes. Searching for a place to worship freely, the first members of the group moved to Germantown, Pennsylvania. A large wave of Swiss Mennonite immigration took place from 1707 to 1757; the first Mennonite settlers arrived in Lancaster County in 1710. It's estimated that prior to the Revolutionary War several hundred Mennonite families settled in Lancaster County.

The migration of these families who separated themselves from the rest of society as well as the rest of the Mennonite populations in the New World represents the first significant reduction in number—referred to as a population bottleneck—which the Mennonites were to experience. Over the next 150 years, this group's numbers grew as a result of reproduction, *not* additional immigration. In other words, the alleles that were available to be inherited came only from those few hundred families.

The nineteenth century proved politically turbulent for the Mennonites of Lancaster County. Mennonite communities in Indiana, Ohio, and Ontario split away from established communities as a result of disputes over evangelism, higher education, missionary work, and conducting services in English. Supporters of the old ways left the larger Mennonite church. In 1893 the Mennonites of Lancaster County experienced a similar schism when Bishop Jonas Martin and followers abandoned the church to establish the Weaverland Conference, an Old Order group.

The Clinic for Special Children's laboratory director Erik G. Puffenberger estimates that 125 families constituted the

Weaverland Conference, and approximately half of the genetic diversity of the conference can be attributed to just thirty families. The split created yet another population bottleneck, and the people who formed the Weaverland Conference weren't a representative sample of the larger population. There are now around 26,500 people who are descended from those 125 families living in Lancaster County and eight other states.

The founder effect can be illustrated with the increase in the frequency of the mutant MSUD allele at each stage of history for the Mennonites. The unusual gene that causes the Mennonite form of MSUD occurs in one out of every 225 people in Europe. When the first group of Mennonites moved to Lancaster County, one person out of a hundred carried the gene as a result of a nonrandom sample of people moving to the county. In the next ten to twelve generations of isolation and intermarriage, the carrier rate for the gene jumped to one out of every thirteen people; this change in allele frequency from one generation to the next is called genetic drift.

Mennonite history provides a fascinating study of the causes of the founder effect. Part of the mission of the Clinic for Special Children is to treat these patients and prevent the devastating neurological damage and death so common with classic MSUD. Building on years of research by other scientists, the Mortons developed a protocol using a special formula lacking the branched-chain amino acids along with a diet high in carbohydrates to treat the Mennonite children who develop the disease. Because children who are sick often metabolize their own tissues, special precautions must be taken during illness to make sure patients don't develop a toxic level of the amino acid leucine. The protocol has proven successful. For example, between 1966 and 1988, 36 percent of the children with MSUD who were lucky enough to make it through infancy died of brain swelling before the age of ten. With the protocol in place, however, from 1988 to 2002 no child under the clinic's care has died from brain swelling.

Like most rare genetic diseases, a cure isn't on the horizon for MSUD patients even though liver transplantation has successfully cured several patients. Through their support of modern medicine, the Old Order Mennonites have provided an understanding of how to treat and live with a once devastating rare genetic disorder. And because of their unique genetic history, in the years to come the Old Order Mennonite and Amish communities will most likely provide answers to more genetic questions.

The Tangier Island Gene

Time hasn't actually forgotten Tangier Island, Virginia. But much of what makes up modern America simply doesn't fit on this windswept outpost—not much more than a sandbar— in the middle of the Chesapeake Bay. The island itself is comprised of three islands that, once settled, were connected by wooden bridges. The main island is the only one currently inhabited because the streams between the other islands have become too wide to maintain the wooden bridges.

It's a place where children walk to school, and automobiles are so uncommon most roads aren't wide enough for them. Bicycles, golf carts, and pedestrians traverse wooden bridges connecting the various parts of the community. Visitors to the community must be ferried over by boat or arrive by air at the tiny general aviation airstrip. The six hundred residents, most of whom are devout Methodists, speak a dialect of English that harkens back to the Elizabethan era when the island was first settled.

For all its isolation and diminutive size, Tangier Island, just one mile wide and three miles long and shrinking, has played crucial roles in both political and scientific history. The launching point for a British invasion of the newly born union and hideout for the occasional pirate also held a closely guarded scientific secret: a rare genetic disorder that ultimately solved a medical riddle about how the body regulates cholesterol.

According to local legend, Tangier Island was discovered in 1608 by Jamestown colony founder John Smith as he was exploring the Chesapeake Bay. The island was first settled by John

Crockett and his sons and their families in 1686 as a place to tend cattle. There are no records to actually confirm the Crocketts were cattlemen, but by 1800 most of the seventy-nine people living on the island carried the Crockett surname and, by that time, farming was their primary occupation.

In 1805 Tangier Island resident Joshua Thomas took on a fateful job ferrying a group of people to hear the Methodist minister Lorenzo Dow preach at a camp meeting. Dow was such a powerful speaker that Thomas was inspired to establish the Methodist Church on Tangier Island.

By the time the war of 1812 broke out, Thomas had become a preacher in his own right. The British arrived on Tangier Island in 1813 and set up camp, complete with breastworks and cannon on the south end of the island near Thomas's meeting grove. The British even intended to build a hospital. Before the British left for their attack on Baltimore, the commanding officer requested Thomas address the disembarking warriors.

Joshua Thomas preached that sermon prophesying doom for the British invaders, an idea that must have provoked some British laughter in 1814 as they sacked and burned Washington, DC, sending President Madison fleeing to the countryside. However, Thomas's prediction proved true at the battle of Baltimore. The Americans withstood the British attack on Fort McHenry, prompting lawyer Francis Scott Key to pen the poem "The Star-Spangled Banner" as he watched the battle.

Tangier Island remained mostly isolated after the war of 1812, and returned to farming. However, in the 1840s the seafood industry became highly profitable, and Tangier Island residents began to drop their agrarian lifestyle and take to the water fishing for crabs and oysters. Today, Tangier Island prides itself as the soft shell crab capital of the world. And, with the exception of tourism, most men on Tangier Island make their living oystering and crabbing.

Tangier Island would have remained a quaint secret to all but the sailors and fishermen on the Chesapeake had it not been for a curious illness in a five-year-old boy. Because Tangier Island is an isolated community, the people living there today trace their heritage to a very few individuals, and as a result people are closely related to each other. Like the Mennonites of Lancaster County, Pennsylvania, this is a recipe for the founder effect.

In 1959, Teddy Laird had just had his huge orange tonsils removed by armed forces physicians; they'd diagnosed him as having a rare disorder known as Neimann-Pick disease and referred him to the National Cancer Institute to see Dr. Louis Avioli. This is where Donald Fredrickson, director of the Molecular Disease Branch came into the picture. Having just reviewed all the literature available at the time for Neimann-Pick disease to be included in a book, Fredrickson knew Avioli's group wasn't looking at a patient with that disease. The boy's tonsils were orange because they were filled with cholesterol-laden cells. In addition to his unusual tonsils, the boy had an enlarged spleen and lymph nodes as well as lacking almost entirely any high density lipoprotein (HDL) or the "good" form of cholesterol, even though the researchers didn't know that at the time.

In 1960, Fredrickson and Avioli headed to Tangier Island to examine residents of the island to see if anyone else suffered from such a condition. It wasn't until they looked at Teddy's sister that they found another set of orange tonsils. That's when Fredrickson discovered the tissues were chock-full of "foam" cells: immune cells called macrophages stuffed to the gills with a slightly modified version of cholesterol. What's more, Fredrickson and his team found these foam cells in bone marrow, liver, spleen, lymph nodes, and other tissues. Their findings sent them back to Tangier Island again to examine the blood concentrations of HDL in the children's parents and relatives. Finding deficiencies in the normal concentration of HDL in the parents, the researchers concluded they were looking at an autosomal recessive genetic defect.

The researchers knew little about HDL and how the disease would progress in adults. As Fredrickson and colleagues continued to follow people who had the disease, they discovered nearly all develop a yellowing of the corneas as they age and peripheral neuropathy in their arms that can prevent the patients from feeling pain. Still, as scientists found more people worldwide who were affected by Tangier disease, they had no idea what the genetic defect in Tangier disease was likely to be.

By the late 1970s, physicians and researchers were aware of a couple of facts. The first was that several proteins, in addition to the cholesterol, comprised the HDL particle. And they knew that people with low levels of HDL were at risk for hardening of the arteries, or atherosclerosis. By the early 1980s, scientists had begun to conclude that HDL was involved in removing excess cholesterol from the body. In other words, HDL worked like a vacuum cleaner sweeping excess cholesterol out of cells.

As a result, many scientists thought they would find that a defect in a major component of HDL—apolipoprotein A1—was the source of Tangier disease. They knew the gene was rare: fewer than seventy-five people worldwide had been diagnosed with the disease, and they knew it behaved as an autosomal recessive disorder. But genome searches throughout the 1980s and 1990s proved fruitless until 1998, when German researchers Gerd Assmann and Stephan Rust, of the Westfälischen Wilhelms-Universität Münster, discovered the gene that caused Tangier disease was found on the long arm of chromosome 9.

Within six months of reporting they had found the region on chromosome 9, Assmann and Rust, and two other teams—one German and one Canadian—had isolated the gene responsible for Tangier disease. The gene they found lends credence to the hypothesis that HDL is a primary actor in removing excess cholesterol from the body: it codes for a member of a group of proteins that transport cholesterol across cell membranes called ATP-binding cassette transporter 1 (ABCA1). Gerd Schmitz and colleagues of the University of Regensberg discovered that five different Tangier disease families harbored five different genetic mutations in the ABCA1 gene. In addition, members of those families who inherited only one copy of the defective gene experienced an "intermediate" condition; their bodies produced about 50 percent less HDL than people with two normal copies of the gene.

Gene in hand, the question then became how do defects in a gene that ferries cholesterol in and out of cells result in the absence of HDL in the blood? Six years later, scientists still aren't entirely sure, but the picture is beginning to resolve. Unlike low density lipoproteins (LDLs, or the "bad" cholesterol), scientists believe HDL particles are formed outside of cells when ABCA1 delivers cholesterol to apolipoprotein A1 (apoA1) to form a brand new, immature HDL particle. Once this nascent HDL particle is formed, up to four additional apoA1 proteins can be added to form a mature HDL particle. In 2005 John Parks of the Wake Forest School of Medicine in Winston-Salem, North Carolina, showed that the liver was the site of most HDL particle formation.

ABCA1 is expressed in a number of cell types, but apoA1 is expressed only in the liver and the intestine. Because all three components of HDL are needed to create a nascent particle, it stands to reason that the liver would be the site of significant HDL formation. The Parks group created a special type of mouse to study the cholesterol problem and found that the ABCA 1 gene was expressed in all the tissues of the mouse except for its liver. In doing so, the team was able to ascertain that 80 percent of the HDL particles were created on the surface of liver cells. The team hypothesized that ABCA1 in the liver may function solely to create nascent HDL particles that would mature and go on to remove cholesterol from other parts of the body using different transport pathways.

Parks team also found that in chickens lacking ABCA1 gene, the kidneys destroyed (catabolized) unbound apoA1 at a much higher rate than kidneys with normal ABCA1. As a result, Tangier disease mutations in the ABCA1 gene disturb cholesterol levels two ways: first, they prevent the creation of nascent HDL particles on the surface of the liver; second, they destroy unbound apoA1 proteins in the kidneys.

But if ABCA1 is expressed in many different cells, why does it act differently between one cell and the other? In July 2005, Hayden's team found the cellular machinery of different cells makes slightly different versions of ABCA1 by choosing to start transcribing the ABCA1 DNA at slightly different places in the beginning of the gene. Other than these different start positions, the ABCA1 proteins are identical to each other. Hayden's team notes the specific ABCA1 RNA transcripts appear at different levels in different tissues.

Their most interesting finding is that after mice are fed a high fat diet, one type of transcript turns on in the liver, while an entirely different transcript comes on in macrophages. As a result, the researchers speculate the transcript predominating in the liver encourages the creation of nascent HDL molecules. In the macrophage, on the other hand, the transcript predominating probably prevents the formation of the foam cells that are so common in Tangier disease.

That opens up a very attractive possibility for therapeutic strategies. Hayden's team has previously shown that ABCA1 expression in macrophages reduce the number of atherosclerotic plaques. Because different transcripts result in ABCA1 expression in macrophages and liver, scientists and drug manufactures could presumably target drugs to stimulate ABCA1 expression in macrophages, while developing an entirely different set of compounds to stimulate the production of ABCA1 protein in the liver to generate a higher HDL level.

To date, there is precious little a person can do to increase their HDL levels aside from diet and exercise; new therapeutic targets offer a substantial amount of hope. The discovery of the Tangier disease gene also offers the opportunity to analyze whether rare alleles associated with Tangier disease play a role in the low HDL cholesterol levels in the general population. Researchers from the University of Texas Southwestern Medical Center and the Copenhagen University Hospital have found that most of the genetics contributing to either very high HDL concentration or very low HDL concentrations are the result of rare mutations in the ABCA1 gene and have enormous effects.

After forty years shrouded in mystery, the Tangier disease is just beginning to reveal its secrets. Perhaps, there is a bit of irony in this story. From an island you can visit only by ferry, we are unlocking the mysteries of a gene critical to ferrying cholesterol out of the body.

The Celtic Curse

A quick scan of the shelves at the local supermarket would lead one to believe Americans are in great peril of suffering from a lack of iron in their diets. The corner store is packed full with iron-fortified cereals, breads, and pastas all in an effort to prevent anemia—the lack of iron in the blood.

Indeed, anemia *is* a serious problem around the world. The World Health Organization (WHO) estimates over half the world's population lacks enough iron in their diets, while a third of the world's population suffers from the weakness and fatigue associated with anemia. WHO experts estimate anemia is a greater cause of ill-health, premature death, and lost wages than hunger.

Faced with the heavy toll anemia exacts, it may seem incongruous that the most common hereditary disorder in North America is an *excess* of iron. Yet, five out of every thousand people in the United States are at risk for developing a potentially fatal overload of iron in their bodies known as hemochromatosis.

Iron acts as the vital core for the oxygen-transporting hemoglobin molecules in red blood cells. Still, too much iron causes just as many problems as too little. Because our bodies make no iron of their own, all of the iron needed for good health must come from the food we eat. Because iron is so critical, the body maintains appropriate iron stores only by regulating the amount of iron absorbed by the intestine. There are no regulated pathways to eliminate excess iron, which means iron can only be eliminated in any real way through blood loss. So, our bodies constantly balance the need to absorb enough iron from the foods we eat against the reality that it has no mechanism for disposing of any excess. When that balance gets out of whack, hemochromatosis or anemia can develop.

First described in the late nineteenth century, patients with hemochromatosis deposit the excess iron in most of their organs, but the iron deposited in their skin—the body's largest organ—leads to a bronzed hue, characteristic of hemochromatosis. While a really good tan may seem like a small consequence, the iron deposits in the liver, pancreas, thyroid, and other internal organs lead to significant medical problems such as liver cirrhosis and cancer, diabetes, and heart failure.

Fortunately, hemochromatosis need not be fatal. The best therapy is familiar to physicians throughout the ages: aggressive bloodletting. For much of recorded history, Western physicians had few tools available to heal the sick. Herbs, poultices, and compresses soothed bruises and cleansed infections. Early doctors could set broken bones and lance abscesses. But when symptoms proved complex, the treatment of choice for most of recorded time was "balancing the humours" by bloodletting.

Bloodletting as a medical practice began in Egypt; Hippocrates documented the practice of curative bleeding around 500 BC, but the practice really took hold in the Middle Ages when a papal decree banned clergy from performing the task but, at the same time, required monks to undergo regular bloodletting. The clergy turned the task over to barbers, who advertised their services with the now-familiar red-and-white pole: red for the blood and white for the tourniquet.

As medical scientists began to understand the underpinnings of human disease, the need to "balance humours" decreased. Bloodletting fell out of favor as it became clear that draining blood usually provided no benefit and, more often than not, seriously compromised patients. One famous example is the death of George Washington in 1799. In order to treat a severe throat infection, physicians repeatedly bled the "Father of the Country" until the former president was so weakened that he died.

It is ironic in this the era of vaccinations, antibiotics, transplant surgeries, and genetic medicine that such a primitive therapy is the best modern medicine has to offer patients with hemochromatosis. Still, the treatment is wildly effective. Patients with iron overloads initially donate a pint of blood a week until their iron stores sink to normal levels. After that, they donate blood once a month to keep their iron levels under control.

What has bedeviled physicians who first see these patients is that too often people with hemochromatosis get their diagnoses *after* iron overload has damaged their bodies irrevocably.

Presumably, identifying patients before their bodies become so overloaded with iron that it causes damage would allow physicians to start phlebotomy (the scientific term for bloodletting) before symptoms even arise.

In 1996, the discovery of a gene related to hereditary hemochromatosis raised hopes that a genetic test could identify people with hemochromatosis early on. Roger Wolff and colleagues at Mercator Genetics in Menlo Park, California, identified a gene on chromosome 6 that was mutated in 85 percent of 178 people diagnosed with hemochromatosis. And of those, all had the exact same mutation in the gene that the group named HFE (*h* for hemochromatosis and *fe*, the scientific designation for iron). The mutation the Mercator team unearthed was a switch in a single nucleotide that caused a guanine to replace an adenine in the HFE DNA. The slight genetic alteration caused the cellular machinery to insert the amino acid cysteine in the spot another amino acid—a tyrosine—usually occupies. That amino acid swap, called C282Y, accounts for nearly all of the mutations in HFE among Caucasians of European descent.

The C282Y mutation proved unusually prevalent in Ireland, Scotland, Wales, and Western France, regions once peopled by the Celts. Scientists believe this mutation arose about six thousand years ago in one individual in a Celtic population. As such, they dubbed hereditary hemochromatosis the Celtic Curse. Researchers speculate that the C282Y mutation became prevalent in Northern Europe as a result of some selective advantage the mutation conferred on the people who carried the gene. Because anemia is such a devastating problem worldwide, many researchers posited that the mutation prevented people from suffering from anemia when dietary iron was in short supply.

With a common mutation in hand and a plausible theory for the existence of that mutation, widespread genetic screening for the disease seems like the next logical step. Such screening would theoretically be fairly straightforward because the mutation conferring a risk for hemochromatosis is the same for most people allowing researchers to develop a one-size-fits-all test. Testing offers physicians a desired means for helping patients *before* hemochromatosis exacts its devastating toll. Just such testing is indeed available; since the discovery of the gene, however, research in the intervening years questions just how much a role mutations in HFE play in the development of iron overload.

Testing for the C282Y mutation revealed that the mutation was far more prevalent than originally anticipated. A 1998 study of 109 randomly selected people working at a hospital in Dublin, Ireland, revealed that 31 of them carried a C282Y mutation in one of their HFE genes. In light of those results, some physicians called for widespread genetic screening to identify patients early.

However, large-scale genetic screening may not be such a useful tool because the C282Y mutation doesn't destine its carriers to disease. The mutation associated with hemochromatosis proved quite common, but the number of people who have that mutation and end up suffering from the symptoms of iron overload is quite low. In 2002 Ernest Beutler and colleagues at the Scripps Institute of Technology in San Diego discovered that most people inheriting two copies of the mutant gene (homozygotes) simply won't develop hemochromatosis at all. (The Scripps study included 41,000 people from the Kaiser Permanente Health Appraisal Clinic.) Beutler and his colleagues compared the symptoms of hemochromatosis, such as liver disease and diabetes, in C282Y homozygotes with the same symptoms in people who lacked the HFE gene mutation. Presumably, more people suffering from these symptoms of iron overload would be C282Y homozygotes. Instead, the team found almost no genetic difference between the two groups studied. The researchers estimated that approximately 1 percent of all homozygotes eventually develop hemochromatosis.

It's not so uncommon for people to carry the genes that should result in a disorder but fail to display the symptoms of the disorder. Scientists refer to the proportion of people with a certain genotype that ultimately displays the phenotype of a disorder as the gene's "penetrance." The Scripps team's results indicate classic hemochromatosis has an extremely low penetrance. A large Norwegian study of 65,238 people supported Beutler's group's findings.

Simply carrying two copies of the C282Y mutation only slightly raises the risk that a person will develop hemochromatosis; therefore, some researchers are beginning to question how valuable widespread genetic testing for hemochromatosis would be.

That's not to say people who carry two copies of the C282Y mutation don't have elevated levels of iron in their blood. They do, but the increased iron levels don't appear to cause the clinical symptoms of hemochromatosis. Whatever role HFE mutations play in the development of the disorder, those mutations alone don't appear to be enough to result in disease. Still, because nearly everyone who develops hemochromatosis is homozygous for HFE mutations, the mutations may help researchers better understand the iron uptake pathway and ultimately find a way to diagnose people before the damage is done.

The HFE protein normally binds to a cell surface protein called the transferrin receptor (TFR). It competes for its spot on the TFR with the iron-carrying protein transferrin. In laboratory studies, this competition results in lower iron uptake by cells. However, HFE protein is strongly expressed by liver macrophages and intestinal crypt cells in the body. Scientists believe crypt cells sense the body's iron requirements, at least in part by monitoring iron uptake by transferrin and programming the amount of iron mature intestinal cells absorb. In patients with hereditary hemochromatosis, these cells behave as though they are iron deficient. In 2004, Hal Drakesmith and colleagues at Weatherall Institute of Molecular Medicine, University of Oxford, discovered that expressing normal HFE in mutant macrophages resulted, paradoxically, in the accumulation of iron in the cells. The English researchers propose that inhibiting the release of iron by these specialized cells plays an important role in regulating iron load in the body and that this function of HFE is lost in people with hemochromatosis.

When a gene displays low penetrance, scientists usually start looking for gene products that interact with it. In 2004 Joanna Poulton and colleagues from the John Radcliffe Hospital in Oxford, England, discovered that mutations in the antimicrobial protein and iron level regulator, hepcidin, can contribute to the severity of hemochromatosis. The team also discovered that a mutation in mitochondrial DNA usually associated with diabetes modulates the severity of iron overload. Given the low penetrance of hemochromatosis, it's likely these are just a few of the modifiers that play a role in whether a person ends up spending time leaking blood at the phlebotomist's office.

Metallic Madness

At the age of twenty-five, the Indian woman's arms and hands had begun to shake uncontrollably. In four months time her legs were moving independently as well. The tremors were bad enough, but then came the mania, which prompted her physician to prescribe potent antipsychotics. And still, she trembled. A neurologist diagnosed her with Parkinson's disease and suggested she undergo surgery to remove the part of her brain that controls movement. At age twenty-nine, she had the brain surgery that excised that portion of her brain called the thalamus, and her tremors decreased. But, one month later the mania returned.

She was referred to the National Institute of Mental Health and Neuro Sciences in Bangalore, India, under the care of a team of physicians led by Dr. A. B. Taly. Among the many tests they ran was an eye examination using a special slit lamp. The patient's irises had a golden brown ring around them. Blood tests confirmed the woman wasn't mad, but she was very, very ill.

Taly's patient, described in the *Journal of Neurology, Neurosurgery and Psychiatry* in 2003, had Wilson's disease—a rare genetic defect in copper transport that had rendered the woman incapable of eliminating the metal from her body. Once she was placed on a copper-reducing regimen, her mania and tremors began to fade.

Wilson's disease causes progressive liver disease and neurological problems. The disorder was first described in 1912 by Samuel Kinnear Wilson. He identified the disease as an inherited condition and dubbed it "hepatolenticular degeneration" because it destroyed the liver, cornea, and brain. It wasn't until 1940 that the cause of that damage was attributed to a deadly accumulation of copper.

Copper is a co-factor for critical enzymes such as superoxide dismutase, an enzyme that sweeps up DNA-damaging free radicals, and lysyl oxidase, an enzyme that crosslinks the structural proteins collagen and elastin. Without a little bit of copper, those enzymes simply don't work. Copper, however, is toxic in large amounts. When it isn't removed in a timely fashion, it is deposited in the patients' brains, liver, and eyes. Accumulation of copper can also cause joint inflammation and kidney damage as well as weakening the heart.

Physicians usually diagnose Wilson's disease in patients who have childhood liver disease. If the child isn't treated or provided a liver transplant, he or she will develop cirrhosis and, ultimately, liver failure. However, if a patient makes it through childhood without showing any symptoms of the disease, neurological symptoms such as the inability to pronounce words, lack of coordination, mania, depression, or psychosis will likely be the first clue that something is amiss. Unfortunately, as evidenced by Dr. Taly's patient, adult symptoms are very often mistaken for other diseases.

Because Wilson's disease is a genetic disorder, researchers were hopeful that discovering the gene responsible for the condition could allow physicians to identify and treat people at risk for the disease *before* their bodies became so overloaded with copper, that they suffered liver and neurological damage. The first clue to the Wilson's disease gene's whereabouts came in 1985 when an Israeli team placed the gene on a region of the long arm of chromosome 13. T. Conrad Gilliam and colleagues from Columbia University narrowed that region to approximately 200,000 base pairs but couldn't get any closer to identifying the gene.

In January 1993, Menkes' disease, also known as kinky hair disease, enters the picture. Menkes' disease is a rare genetic neurodegenerative disease that causes mental retardation, seizures, low body temperature, loose skin, arterial rupture, twisted hair, and early childhood death. Whereas Wilson's disease arises from too much copper, people with Menkes' disease suffer from fatal copper insufficiency. Patients with Menkes' disease absorb copper from their diets, but the copper never makes it out of the cells lining the gastrointestinal tract. Three different groups from Britain, Michigan, and San Francisco isolated the gene for Menkes' disease on the X chromosome and discovered it coded for a copper transport protein known as an ATPase, specifically ATP7A.

The ATP7A gene is expressed many cell types presumably to aid in copper transport, but it's never expressed in the liver. Because copper can't be transported out of the liver in Wilson's disease, scientists suspected the responsible gene would code for a protein similar to ATP7A but one that was expressed and active in the liver. Gilliam then collaborated with Rudolph Tanzi, a neurologist at Massachusetts General Hospital who had been looking for metal-binding proteins as part of his work on Alzheimer's disease and discovered a metal-binding gene on chromosome 13. The two researchers realized in 1993 that what they had found was another copper transporting ATPase—ATP7B—the cause of Wilson's disease. At the same time, Diane Cox's group at the Hospital for Sick Children in Toronto also identified the gene as the cause for Wilson's disease.

In contrast to the expression pattern of ATP7A, ATP7B is expressed predominantly in the liver, kidney, and placenta. So the two disease genes perform the same function—transporting copper—but the diseases are vastly different: in Menkes' disease the defect affects a tissue where the transporter is designed to bring copper *into* the body, and in Wilson's disease the fault lies in a protein that is supposed to help the liver take copper *out of* the body.

Armed with the gene, Tanzi and Gilliam's group also reported that one particular mutation was responsible for up to one third of all the Wilson's disease cases in Europeans. When a certain mutation commonly causes disease, it raises scientific hopes that a simple DNA screening test is just around the corner. Despite the early excitement, developing a simple screening test for

Wilson's disease hasn't been an easy task. There are currently over two hundred known mutations in the ATP7B gene that lead to Wilson's disease. Still, scientists around the globe are analyzing these mutations looking for combinations of mutations that lead to the variations that can be seen in Wilson's disease phenotypes.

Without a genetic test, treatment once symptoms are identified is the standard of care. Several agents such as penicillamine and ammonium tetrathiomolybdate bind accumulated copper so that it can be excreted. And, because all patients with neurological symptoms always develop brown copper deposits on the rims of cornea known as a Kayser-Fleischer ring, physicians suspicious of Wilson's disease could employ a simple slit-lamp examination to identify people suffering from the metallic madness. But, of course, they first have to suspect a very rare illness in order to perform that very simple test.

chapter Two

Just One Bad Apple . . .

Your third-grade teacher was right. Sometimes it takes only one bad apple to spoil the barrel. In other words, for some genetic disorders, just one faulty gene serves as the trigger. Inheriting it from either your mother or your father pretty much ensures you will have the condition yourself. This is termed "autosomal dominant inheritance"—the word *autosomal* simply refers to the fact that the gene for whatever condition doesn't reside on either of the sex chromosomes. The word *dominant* means inheriting one gene is enough. For example, if your father had an extra finger or toe next to his pinky finger or toe, you and your brothers and sisters each would have a 50 percent chance of having an extra digit as well.

When a gene produces a dominant trait such as an extra digit, the inheritance pattern displays some typical features. The first was already mentioned; each child of an affected parent has a 50 percent chance of developing the disorder. In addition, the condition doesn't skip generations. If your father is, in the words of the *Princess Bride*'s Inigio Montoya, a "six-fingered man," and none of his children have six fingers or toes, then none of his grandchildren or great-grandchildren will inherit the trait from him. Autosomal dominant traits affect males and females in roughly equal numbers.

Autosomal dominant inheritance is the most straightforward inheritance scheme in Mendelian genetics. Yet even among autosomal dominant diseases there can be some complexity. For example, sometimes more than one gene can lead to the same phenotype as is the case for familial hypercholesterolemia, a condition that can cause very high cholesterol levels. What's more, other genes and environmental influences can also play a role in determining when or how severely someone is affected with a dominantly inherited disease. Dominant genes provide the closest genetic example of fate, but it may be fate with an asterisk.

The Long Stretch Gene

On January 24, 1986, U.S. Olympic volleyball player Flo Hyman took a well-earned breather during a game her team was playing in Matsue, Japan. It was the third game of the evening, and Hyman rotated out on a routine substitution. She sat on the bench and within seconds slid to the floor. Just two years after her team made history winning a silver medal in Los Angeles, the woman touted as the best female volleyball player ever was dead.

Within an hour, physicians attributed her death to a heart attack. Her family, however, wasn't convinced that a healthy athlete's heart would simply give out, and they ordered an autopsy. Her family had been right; Hyman's heart had been strong. The autopsy found that the major blood vessel leading away from it, however, was weak, and when it ruptured Hyman was dead in moments. As shocking as the death of this vital thirty-one-year-old was the cause: Flo Hyman had the genetic disorder Marfan syndrome.

Looking back, the towering six-foot-five Hyman had many of the outward signs of Marfan syndrome. Her height and slender build are characteristic of people with Marfan. Her long angular face and unusually long arms, too, were visible harbingers, as were the long, slender hands. Even so, those characteristics weren't so unusual that her coaches or doctors noticed. Looking at her face you were likely to notice lively eyes and an electrifying smile. None of the other manifestations of Marfan syndrome particularly stand out among elite volleyball players who tend to be tall, lean, and long-limbed.

But it's the less visible characteristics that are of particular consequence to the one in twenty thousand people in North America who have Marfan syndrome. In addition to pronounced lankiness, people with Marfan syndrome can have a laundry list of abnormalities, but it is the heart and blood vessel defects that are most lethal because they can lead to congestive heart failure and, as in the case of Flo Hyman, aortic rupture.

The problem for patients with Marfan syndrome involves connective tissue. This tissue serves as the structure and glue for the body, and includes such things as cartilage, ligaments, and the anchoring substance around cells called the extracellular matrix. In patients with Marfan syndrome, a critical protein used to build all connective tissue is defective, and as a result their connective tissues are just too stretchy.

Marfan syndrome was first described in 1896 by the French physician Bernard Marfan. He noticed a young girl whose arms, legs, fingers, and toes were long and thin compared to her torso. The girl also suffered from a curvature of her spine and poor muscle development. Shortly after Marfan identified the syndrome, physicians began to identify other characteristics of the condition. In addition to the symptoms already mentioned, patients may have a caved in or domed breastbone, overly flexible joints, flat feet, and a high, arched palate that causes their teeth to be crowded.

More than half of all Marfan syndrome patients end up with a dislocated lens of the eye, which can be slight or quite noticeable. People with Marfan syndrome also tend to be nearsighted and are at risk for early glaucoma and cataracts.

Most Marfan patients suffer not only from a weakened aorta but also from a mitral valve prolapse—where the valve between the upper and lower chambers on the left side of the heart billows out and causes blood to back flow from the lower chamber into the upper chamber. Mitral valve prolapse can be especially painful during times of stress. More serious is when the aorta widens where it attaches to the heart. As the aorta widens, the left ventricle—the lower chamber of the heart and its major pump—enlarges as it must pump even harder. Ultimately, the excess pumping force damages the heart muscle so severely that the patient develops congestive heart failure.

Scientists first began studying connective tissue in earnest in the 1950s, and Marfan syndrome was an obvious choice for analysis because the symptoms were so disparate but clearly related to connective tissue. Scientists were baffled by what possible defect was responsible for the changes in the aorta as well as the lens of the eye.

The first clue came from Lynn Sakai and colleagues at the University of Oregon Health Sciences Center. In 1986, the team discovered a protein they called fibrillin, which serves as the key component of a microfibril. Microfibrils are minute rod-like structures that can, in some tissues such as the aorta and ligaments of the musculoskeletal system, provide the framework for elastic fibers. In other tissues, such as the filaments that hold the lens in place in the eye, they have no association with elastic fibers at all. The common link between these two tissues is the fact that microfibrils have a remarkably uniform structure from one tissue to the next.

In 1991, a group of researchers from Oregon, Baltimore, and Boston—each taking a different tack—honed in on the gene for fibrillin and simultaneously announced defects in the gene located on chromosome 15 were responsible for Marfan's syndrome.

Marfan syndrome displays a typical characteristic of many single gene diseases: a single defect can have many effects on different areas and systems in the body. Geneticists call this "pleiotropy." In the case of Marfan syndrome, the genetic defect readily explains why both the aorta and the lens of the eye are affected. It even explains how the limbs become so long: bone is covered by a connective tissue called the periosteum. Among other contributions to bone health, the periosteum pro-

The portion of a gene that codes for a protein isn't always one contiguous block on a chromosome. Often the gene is made up of coding portions called exons, and noncoding portions called introns.

vides an oppositional force to bone growth. When the periosteum is too stretchy, bone overgrows.

When the fibrillin gene was identified as the cause of Marfan syndrome, there was great hope in the medical community that it would lead not only to new therapies but also permit them to easily identify people with Marfan using a simple genetic test. Too many Marfan patients are diagnosed only after they've died from an aortic rupture, as Flo Hyman was. Unfortunately, creating that test hasn't turned out to be so simple.

The fibrillin protein is large, and the gene encoding it even larger. The gene itself is more than two hundred kilobases in length, and the portion of the gene that codes for the protein is divided up into sixty-five different chunks called exons. The

When the cellular machinery transcribes a gene comprised of introns and exons, the result is an RNA that needs a little processing. A different set of cellular proteins splices the exons together so the coding portions are now contiguous and ready to be translated into proteins.

protein can only be made once the cellular machinery pieces those chunks together in the form of one contiguous strand of RNA. As a result, there are many opportunities for a mutation to occur. To date, there are nearly two hundred different mutations associated with Marfan syndrome, making it impossible to produce a simple genetic test for the disorder.

Still, researchers have been able to make efforts to use the genetic information to better understand the disease. For example, researchers have been correlating the genotypes of the mutations with the phenotypes presented in the disease. In general, missense mutations (changes to individual DNA base pairs resulting in the protein being made with a different amino acid) have the most serious effects in Marfan syndrome. Harry Dietz, a Howard Hughes Medical Institute scientist at Johns Hopkins University in Baltimore, Maryland, and colleagues found that this happens because the mutated fibrillin proteins bind to and disable the normal fibrillin. In other words, plenty of fibrillin is being made from the normal copy of the gene, but the protein made from the defective copy prevents the normal fibrillin from being deposited in the connective tissue the way it's supposed to be. When a mutant protein actually prevents a normal protein from functioning, geneticists call it a "dominant negative effect."

Because one defective gene causes this disease, scientists had been searching to find out what happens when a person inherits two defective copies of the fibrillin gene. Swedish researchers, led by Karl-Henrik Gustavson at the Medical Center Hospital in Orebro, Sweden, identified just such an anomaly in a newborn boy. The boy had inherited one mutant copy of the gene from his mother and a different mutant copy from his father. Unfortunately, that inheritance pattern proved devastating: the child was born with severe congestive heart failure and died by the age of four months.

Despite research into which mutations prove most deleterious, physicians still can't predict which cases of Marfan syndrome will be the most severe based solely on genetic analysis. As a highly variable disease with a markedly visible phenotype, researchers and historians have speculated whether the condition has affected unusually tall individuals over the course of history. Historians have suggested that composer and pianist Sergei Rachmaninoff may have suffered from the disease but presumably benefited from an increased hand reach. Violinist Niccolo Paganini and Mary Queen of Scots have also been mentioned as potential Marfan sufferers. In the United States, the sixteenth president, Abraham Lincoln, has garnered the most questioning.

Lincoln was by all accounts a tall man. Exactly how tall no one knows, but as Phillip Reilly notes in his book *Abraham Lincoln's DNA*, the photos of the president at the Antietam battlefield show him towering over all of the assembled military officers. Even so, when he was sitting Lincoln didn't seem much taller than other men—most of his length was in his arms and legs. In addition, Lincoln was known to have astonishingly large feet as well.

The idea that Lincoln had Marfan syndrome took root after California physician Harold Schwartz diagnosed a boy with the condition and discovered the child was a descendent of Lincoln's great-great-grandfather Mordecai Lincoln. Schwartz published his speculation in 1962 in the *Journal of the American Medical Association*.

Schwartz's hypothesis has been debated in the intervening decades, but when the gene for Marfan syndrome was discovered, suddenly there was the possibility that scientists could put the debate to rest. Their challenge now was to obtain something that contained Lincoln's DNA. The National Museum of Health and Medicine in Washington, DC actually has such items in its collection: locks of hair, blood-stained clothes from the night Lincoln was assassinated, and other items. Extracting the DNA, however, would mean destroying at least a portion of whatever sample any researcher obtained.

Geneticist Darwin Prockop, then at the Jefferson Medical College in Philadelphia, jumped on the opportunity and requested permission in 1991 to obtain DNA from some of those stored samples in order to test the DNA for Marfan syndrome. The museum convened a panel of science and ethics luminaries (including Reilly, and Johns Hopkins University geneticist and Marfan expert Victor McKusick, who had written extensively on Marfan syndrome) to consider the ethics and practicalities of the request. In May 1991 the panel delivered a "qualified green light" for the proposal based on privacy, ethics, and legal considerations. The question remained whether it was scientifically feasible to conduct such a test on valuable artifacts. In April 1992 the panel decided that it was not time to proceed with such testing because a simple genetic test didn't actually exist, and the entire gene would need to be studied to make any determinations. In addition, the panel felt DNA extraction techniques weren't adequately advanced to guarantee the maximal amount of DNA would be recovered from any sample.

The request for Lincoln's DNA was one of the first attempts to reach back in history and use genetic information to answer nagging questions. But it certainly hasn't been the last. Genetic analysis has been used on remains to determine if they were actually those of Butch Cassidy, Pizarro, and Czar Nicholas of Russia. As genetic techniques continue to improve, the requests are likely only to multiply. Lori Andrews of the Illinois Institute of Technology and colleagues maintain that while the interest in such information is understandable, some ethical guidelines need to be in place for such inquiries.

At the same time, when reports of genetic diseases focus so heavily on the disability, patients with those conditions find great solace in learning important historical figures in some part shared their struggles and fate. For now, Marfan syndrome patients need only to look to Flo Hyman's accomplishments for certain inspiration. Lincoln's story may be answered later.

The Dracula Gene

At night, the colorfully striped zebra fish looked like all the other zebra fish that had been captured from freshwater streams in India to populate American fish tanks. But this fish looked much different during the day.

Just like Count Dracula, this particular fish abhors the sunlight. When exposed to light, the animal's red blood cells turn fluorescent and simply explode. This poor blighted fish suffers the same malady that felled a British monarch. Like King George III, this zebra fish has a defect in the pathway that produces the oxygen-carrying molecule heme.

Zebra fish have become one of the favorite experimental animals among developmental biologists studying vertebrates because their embryos are transparent. As a result, the researchers can watch the embryos develop all the major structures of their nervous system from a sheet of cells to the neural plate to a fully structured brain. In addition, scientists can watch the very earliest stages of embryogenesis.

Our poor vampiric mutant zebra fish survives embryogenesis just fine; it just can't make stable red blood cells. When researchers at the Massachusetts General Hospital in Boston identified the gene that had gone awry, they designated it "dracula." It codes for an enzyme called ferrochelatase and is a model for the human disease erythropoietic protoporphyria.

Erythropoietic protoporphyria (EPP) is a rare, largely autosomal dominant condition; there are a few families where the disorder appears to follow a recessive inheritance pattern, but most cases result from simply inheriting a single mutated copy of the ferrochelatase gene. EPP is one of seven conditions known as porphyrias that hamper the body's ability to synthesize heme, the molecule that delivers oxygen to the body. Ferrochelatase is the final step in the production of heme. As a result, the heme metabolite Protoporphyrin IX builds up in the blood stream.

That buildup of Protoporphyrin IX causes patients with EPP to have extremely sun-sensitive skin. Sun-exposed areas of the skin become itchy, burn, and sting for as long as hours after stepping out into the sun. Sometimes the skin becomes red and swollen, but it usually heals without scarring. Most often, symptoms begin in childhood. Compared to other porphyrias, however, the symptoms of EPP are relatively benign. They are certainly minor in comparison to the zebra fish carrying the same mutation.

The porphyrias, however, aren't uniformly mild or easy to identify. Often times they mimic other diseases or present such a bizarre panoply of symptoms that doctors are at a loss to identify what causes the trouble. For example, just one step prior to ferrochelatase in the heme synthetic pathway requires the work of an enzyme called protoporphyrinogen oxidase. A defect in the gene encoding this enzyme (PPOX) results in variegate porphyria, another autosomal dominant disorder.

As its name implies, variability is the salient feature of this type of porphyria. For as many as 80 percent of all those who harbor mutations in the PPOX gene, they will never know anything is wrong. Others, however, may suffer a series of acute "attacks" characterized by abdominal pain, constipation, diarrhea, vomiting muscle weakness, seizures, hallucinations, and delirium. People suffering variegate porphyria also suffer from photosensitivity to a degree that their skin blisters and scars in response to sunlight. In addition, during an attack, a porphyric's urine also will turn dark red or purple upon exposure to light and air.

While people who harbor this gene may live for decades without ever having an attack, they are always at risk if they're exposed to the right pharmaceutical or environmental triggers. Barbituates, sedatives, certain antiseizure medicines, anesthetics, infections, severe dieting, and estrogen can all trigger porphyric attacks. Treatment includes a high carbohydrate diet and prompt treatment of infections as well as avoiding triggering substances.

Variegate porphyria is unusually common among South African descendents of a seventeenth-century Dutch settler Gerrit Jansz and his wife, Ariaantje Jacobs. In South Africa, three in one thousand people will have porphyria. The incidence is so high that physicians take special precautions when prescribing medicines and using general anesthesia by routinely testing for porphyria anyone heading into surgery.

In 1969, Ida Macalpine and her son Richard Hunter—two British psychiatrists—reviewed the medical records of King George III, the British monarch at the time of the American Revolution. During George III's reign, Britain gained supremacy over the seas, defeated Napoleon, and became a colonial power. At the same time, George's time on the throne was blemished with repeated episodes of mental instability as well as the loss of the American colonies. That loss may be directly attributable to madness induced by porphyria.

King George III suffered five episodes of protracted and severe derangement. The first attack came in 1765. At various times during his reign the king was confined, straight-jacketed, and institutionalized as result of his insanity. During his illnesses, his physicians noted he passed purple urine and suffered from profuse sweating, skin lesions, rapid pulse, delusions, and delirium—all symptoms of variegate porphyria. George III died blind, deaf, and mad at Windsor Castle on January 29, 1820.

George III's illness didn't simply appear out of the blue. Macalpine and Hunter noted that Mary Queen of Scots, one of George III's ancestors, is thought to have suffered from porphyria as she was known to have suffered from intermittent "colics," paralysis, and madness. Her attempts to assassinate her cousin, Queen Elizabeth I, could well have been simply the outcroppings of her deluded mind. King James VI, Queen Mary's son, is also likely to have suffered from porphyria because he claimed to have urine the color of wine.

The case for King George III suffering from porphyria is compelling but with one caveat: his attacks were unusually long, severe, and persistent. After all, 80 percent of all porphyrics never have an attack at all. That's not to say an attack of porphyria is benign; it can be deadly. But for people who suffer from attacks, they often find subsequent ones less severe. Not so for George III.

An analysis of the king's hair (reported in 2005) may explain just what made his attacks so unusual: high exposure to arsenic. Knowing that porphyria attacks can be the result of exposure to heavy metals such as lead and mercury—metals found in hair powder and makeup used at the time—Martin Warren and his colleagues at the University of Kent examined a sample of the king's hair. Rather than finding the metals they suspected, Warren's team instead discovered the king's hair contained arsenic at the extremely high level of seventeen parts per million, a great deal more than the less than one part per million most people enjoy.

Arsenic interferes with the production of heme. For a patient who already has difficulties with heme metabolites building up in his blood stream, any further perturbation of the heme biosynthesis pathway will only compound matters. As a result, exposure to arsenic often triggers attacks of pain and madness for patients with porphyria.

But where was the king getting such exposure to arsenic? Warren thinks he knows. A review of the monarch's medical records indicates George III was consistently administered emetic tartar—an antimony-based medicine—by his physicians to treat his severe abdominal pains. The substance does have some medical value. For example, to this day, the semimetallic element antimony is used to treat the parasitic infection leishmaniasis. In the eighteenth century though, antimony was simply dug up from the ground and subjected to precious little purification. Antimony is often found naturally in areas that have high levels of arsenic. Warren believes the medicines with which the royal physicians so liberally dosed King George III were the source of the arsenic in his hair, and quite possibly the cause of many of his bouts with madness.

The king's madness played out on an international stage during some of the most important moments in the birth of the United States. During his first illness, relations with the colonists in North America became markedly strained. We don't know if George III's madness played any role in precipitating the Revolutionary War, but medical historians will continue to argue whether George III actually had porphyria and to what degree the disease itself played in increasing the tensions between Britain and its colonies.

The Expandable Gene

Unusual grimacing is a first sign that something might be amiss. Soon afterward absentmindedness sets in, and gestures start to become quite out of the person's control. Involuntary dancelike movements can become so frequent and severe that the uninformed may think the person is drunk. Planning even the simplest tasks becomes almost impossible, and as Huntington's disease runs its course, depression, aggressiveness, even dementia and psychosis take over the mind.

This grim picture is made worse by the fact that if it is your parent succumbing to the illness, you have a 50 percent chance of developing it yourself. Should it strike you, it will most likely do so at an even earlier age. As genetic diseases go, Huntington's is devastatingly close to the epitome of genetic determinism. Inheriting a mutated gene means you will develop Huntington's disease. Nevertheless, as researchers continue to learn more about the gene that causes this disease, they are finding even something that seems so fateful may be able to be modified.

Huntington's disease is named for George Huntington, a physician whose father and grandfather were also doctors on Long Island, New York. He first encountered the disorder when he was a young boy accompanying his father on professional rounds. In 1872, the twenty-two-year-old published his observations in *The Medical and Surgical Reporter of Philadelphia*. Noting the grotesque uncoordinated movements, he termed the disease a chorea—the Greek word for dance—and described the symptoms of the disease as well as its characteristic onset in middle age. In addition, he accurately described the inheritance of Huntington's disease noting that "if by any chance these children go through life without it, the thread is broken and the grandchildren and great-grandchildren . . . may rest assured that they are free from disease."

Huntington's disease is an autosomal dominant genetic disorder that strikes approximately one in ten thousand people in the Americas, Europe, and Australasia. It is significantly less common among the Japanese and Africans. People usually show their first symptom between the ages of thirty and fifty; however, the disease has been observed as early as age two.

The progressive loss of motor control in Huntington's disease coincides with the loss of neurons primarily in the brain's cortex and striatum. Some patients lose as much as 25 percent of their total brain weight as a result of this disease. Its course is inexorably, horrifyingly slow, leading to total disability and eventually death fifteen to twenty years after diagnosis.

In 1968, when she was twenty-three years old, Nancy Wexler learned the devastating news that her mother, Leonore, had Huntington's disease just as did three of her uncles and her grandfather. The personal toll the disease took on her family spurred the entire family to devote themselves to finding a cure for Huntington's disease. After all, she and her sister, Alice, had a 50-percent chance of succumbing to the disease themselves.

After obtaining her PhD in psychology from the University of Michigan, Wexler began her work on Huntington's disease. This took her, in 1981, to a remote village on the shores of Lake Maracaibo, Venezuela, where Huntington's disease occurs in near epidemic proportions. First identified by Amerigo Negrette in the 1950s, the Lake Maracaibo community has been and continues to be critical to our understanding of Huntington's disease. Wexler and her colleagues developed a Huntington's disease pedigree for the Venezuelan villagers that reaches back ten generations and contains more than eighteen thousand individuals. Nearly fifteen thousand of them trace their origin back to one woman with Huntington's disease, Maria Concepción Sota, who lived in a stilt house on the shores of the lake in the early nineteenth century.

In 1981, Wexler's mission was to build a pedigree and draw blood that would be used to search for a genetic marker for Huntington's. Working with Jim Gusella of the Massachusetts General Hospital, she sent the pedigrees and the blood to Gusella for analysis. In 1983, Gusella found that marker on the tip of the short arm of chromosome 4.

The tip of a chromosome is a slippery place to work, however. Finding the actual gene for Huntington's disease took ten more years of work by dozens of investigators before it could be cloned. The gene they found is huge—it contains 67 exons spanning a total of 170,000 DNA base pairs. Once the gene has been processed it codes for a 3,144 amino acid protein, which the researchers dubbed "huntingtin."

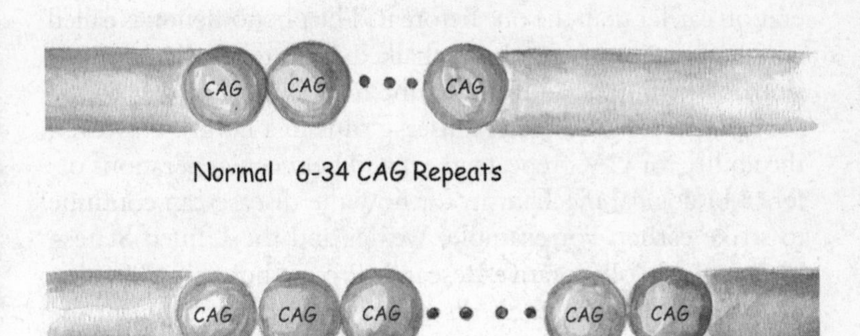

The huntingtin gene contains a region where a triplet of nucleotides, CAG, is repeated many times. The normal gene carries anywhere between 6 and 34 triplet nucleotide repeats. For people suffering from Huntington's disease, those repeats have increased from as few as 40 repeats to as many as 180 repeats.

Curiously, at the very beginning of the gene is a string of repeated nucleotides: CAG, CAG, CAG, and so on. This triplet-repeat encodes many iterations of the amino acid glutamine and is referred to as a polyglutamine tract. In normal people, the number of CAG's in this stutter is stable. But in people with Huntington's disease, the number of CAG repeats in the gene can become especially prolonged. Normal individuals carry huntingtin alleles with between 6 and 34 CAG repeats while people destined to develop Huntington's disease possess at least one allele containing between 40 and 180 CAG repeats. For those with 35 to 39 CAG repeats, some will develop Huntington's disease and others won't.

For unknown reasons the number of CAG repeats can increase from generation to generation, especially when the mutation is inherited from the father. But even more interesting, the greater the number of CAG repeats, the earlier Huntington's disease will strike.

Physicians and Huntington's disease families have long maintained that Huntington's disease strikes each successive generation earlier than the one before it. The phenomenon is called anticipation, and many have chalked it up to medical professionals looking for and finding the telltale signs of the disease sooner when Huntington's disease runs in a family. However, the ability for CAG repeats to expand between generations offers a biological mechanism for how the disease can continue to strike earlier. For example, Wexler and the United States–Venezuelan Collaborative Research Project noted in 2004 that in "typical" Huntington's disease, which strikes between the ages of twenty-one and fifty, the median number of repeated CAG sequences is 45. Whereas, in the juvenile form of Huntington's disease, which strikes between the ages of two and twenty, the median number of CAG repeats is 60.

Huntington's disease is almost certainly *not* the result of the loss of the normal function of huntingtin protein because there is presumably a normal copy of the huntingtin gene available to provide the cell with the protein's normal function. In addition, people suffering from Wolf-Hirschhorn syndrome—a rare condition where a large chunk of chromosome 4 is missing, including the part containing a copy of the huntingtin gene—show no evidence of any Huntington's disease symptoms.

Instead, it is more likely that the extra CAG repeats result in the mutant huntingtin protein *gaining* some function over and above that of normal huntingtin. It's been more than ten years since the huntingtin gene was discovered; still, researchers aren't quite sure what it normally does in the cell. Various researchers have suggested it plays roles in such cellular processes as growing neurons, programming cell death, and moving proteins in and out of cells. What they do know is that it is a protein misfolding disease. Mutated huntingtin proteins build up as toxic aggregates in and around the nuclei of nerve cells causing neuronal death. Because extended tracts of the amino acid glutamine arrange themselves in space as sheets, they create a large sticky surface that likely facilitates aggregation between huntingtin as well as other proteins in the cell.

Even though the most critical determinant of when Huntington's disease will strike is the number of CAG repeats, work by Michael Hayden, of the University of British Columbia, and Peter Harper, of the Institute of Medical Genetics, University of Wales College of Medicine, indicated other modifying factors existed because individuals with the exact same number of CAG repeat have significant variation in their age of onset with the disease. Wexler and the United States–Venezuela Collaborative research project examined the Venezuelan kindred to identify the extent of modifying factors.

The Venezuelan patients on average experience an earlier age of onset compared to North Americans. The team conducted both genetic (number of CAG repeats) and phenotypic (age of onset of first Huntington's disease related motor symptom) analysis of the Venezuelan group. The researchers examined the number of CAG repeats in the mutant copy of Huntington's disease and compared it to the number of CAG repeats on the normal allele. Because Huntington's disease is so prevalent among the Venezuelans, some of them had either two mutant alleles or a mutant allele and one in the 35–39 repeat range. In all cases, the researchers established age of onset with the longer allele and examined what role the shorter allele played in determining when the person developed Huntington's disease. Surprisingly, they found the shorter allele had no effect at all.

Instead, by comparing groups of siblings to each other, they found 84 percent of the variation in age of onset could be attributed to shared genes other than huntingtin and also shared environmental factors. Wexler notes that such a discovery means that identifying those genes affecting age of onset offers another target for developing pharmaceuticals that could postpone the onset of disease, even if it couldn't prevent the disease all together.

The end result of such a strategy could mean many additional years of productivity for people with Huntington's. In addition, it offers many more years of hope that science will unearth what role huntingtin protein plays in the brain and find a way to prevent the disease entirely. What Wexler's work shows us is that even when the disease is a certainty, some things remain uncertain.

chapter Three

You Can Blame It on Mom

Ever since Freud put the focus on experiences in early childhood, everything from bad manners to sociopathy has been blamed on mom. The jury is still out on whether mothers wield so much psychological power over their offspring that every sort of bad behavior can be linked directly back to them. However, because of the peculiarities of how sex is determined, sometimes the only parent who matters is mom.

The human genome comprises twenty-three pairs of chromosomes; only one pair of those chromosomes determines sex. We are born male or female depending on which combination of sex chromosomes we inherit. Inheriting a matched set of two X chromosomes results in a female; an X and a Y combination produces a male. Because mom has only X chromosomes to donate, every son and daughter receives one X chromosome from mom. Dad donates the complement to mom's X chromosome: to sons it's a Y chromosome, and to daughters it's another X chromosome. When a gene on the X chromosome goes awry, mom is almost always the source of the trouble, and sons are most at risk.

A quick look at the numbers illustrates why. When females inherit a faulty X-linked gene from one parent, they have a pretty good chance of inheriting a normal copy on the other X chromosome. Males, on the other hand, have only one chance at inheriting normal X-linked genes . . . and their luck depends entirely on mom.

The Gene That Launched a Revolution

Genetic diseases often wreak havoc on individuals and their families. Even so, it's rare that a genetic disease has any impact on completely unrelated members of society not to mention instigating the overthrow of a government. But that's exactly what happened when the heir to the Russian throne inherited the "royal disease"—hemophilia.

The propensity for the male children of certain families to suffer severe and frequently fatal bleeding episodes has been documented since ancient times. In the second century AD, the Talmudic scholar Rabbi Simon ben Gamaliel of Jerusalem excused from ritual circumcision the sons of women who'd had two other sons die from excessive bleeding. In the twelfth century, the Hebrew physician and philosopher Moses Maimonides extended that exemption from circumcision as a result of bleeding to half-brothers with the same mother. The Arab physician Albucasis wrote in the twelfth century of a family whose sons bled to death after suffering minor injuries.

Centuries before Mendel experimented with his peas, these ancient scholars recognized a key point about hemophilia—it doesn't play fair. Men get it, but only women pass it on. And arguably the most famous mother to pass on hemophilia was Britain's Queen Victoria.

Hemophilia is the quintessential X-linked recessive disease. The genetic defect responsible for hemophilia resides on the X chromosome. Because women have two X chromosomes and men only have one, all it takes for a boy to develop hemophilia is to inherit a faulty X chromosome from his mother. In order for a girl to develop hemophilia, she would have to inherit one faulty X from her mother and one also from her father. Such a situation is possible but far less likely. Each daughter of a carrier mother has a 50 percent chance of being a carrier herself. But, every daughter of a hemophiliac is a carrier of the disease.

Originally, physicians erroneously suspected that structural defects in blood vessels caused hemophilia. Instead, research throughout the middle years of the twentieth century pegged hemophilia as a clotting problem. In 1937, Patek and Taylor found in laboratory experiments that they could correct the clotting problem in hemophiliacs' blood with a substance in plasma. In 1944, an Argentinean physician named Pavlovsky described a case from a laboratory clotting experiment where the blood from one hemophiliac corrected the clotting defect in the blood from another hemophiliac and vice versa. This experiment gave the first clue that hemophilia wasn't one single disease. Pavlovsky's findings weren't fully explained until 1952 when University of Oxford physicians Richard Macfarlane and Rosemary Biggs described Christmas disease (also known as hemophilia B)—a distinct form of hemophilia named after Stephen Christmas, a ten-year-old boy suffering from the disorder.

X-linked hemophilia exists as two separate diseases, hemophilia A and hemophilia B, which are indistinguishable without laboratory or genetic tests. The discovery that they were separate disorders laid the ground work for understanding how the body stopped bleeding. When a person suffers a cut, their body employs a cascade of coagulation proteins that scientists refer to as factors to staunch the bleeding. In 1962, an International Congress named each of these proteins and their associated bleeding disorders. A defect in the coagulation protein Factor VIII was responsible for hemophilia A, and a defect in Factor IX was responsible for hemophilia B. In 1964, Macfarlane published a detailed description of the coagulation cascade in the journal *Nature*.

With the cascade in hand, scientists could begin the process of developing purified plasma extracts of the different coagulation proteins to deliver a means to treat and prevent severe bleeding episodes. Researchers were also able to begin answering such questions as how many people get these diseases? What is the effect on carrier women? What proportion of patients has severe vs. mild disease?

In the United States, four hundred hemophiliacs are born each year. Approximately 85 percent of those have hemophilia A, and 15 percent have hemophilia B. Nearly 70 percent of hemophilia A patients are classified as severe bleeders—meaning they have less than 1 percent of the Factor VIII activity of nonhemophiliacs. Patients with hemophilia B, on the other hand, are more likely to have mild disease.

An interesting situation arises for women carrying one copy of the hemophilia A gene defect: they produce about 50 percent less Factor VIII than noncarriers. As a result, the carrier status affects overall health, although not in an entirely beneficial manner. On the plus side, carriers of hemophilia A are 36 percent less likely to die from heart disease caused by hardening of the arteries. These same women, however, are more likely to die from brain hemorrhages, so it's unlikely that carrier status is a net benefit.

At the same time scientists were delineating the coagulation cascade, they were beginning to map genetic defects associated with clotting problems. From the early 1960s researchers scanned the X chromosome, painstakingly plotting the defects' distance from known genes and chromosome structures . . . such as the waistlike center called the centromere and the very tip ends known as telomeres.

In 1982, the smaller protein responsible for hemophilia B, Factor IX, was the first to be identified and cloned by a group led by the University of Oxford's George Brownlee. Two years

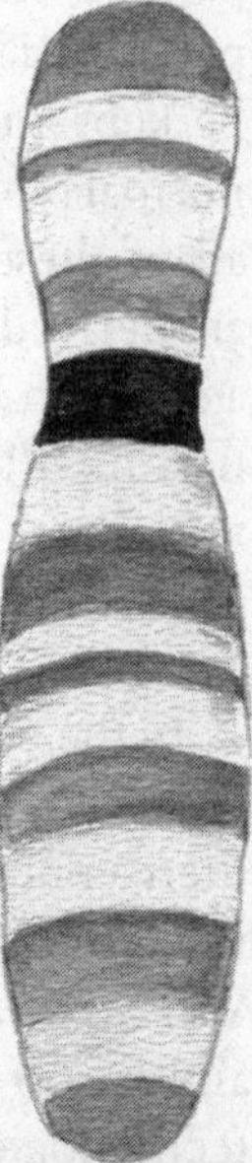

Every chromosome has a centromere, which can be seen as an indentation on the chromosome when a cell replicates. The centromere serves as an anchor for the cellular scaffolding that divides a replicating cell and as a reference point for geneticists.

later, the enormous Factor VIII—a gene with 26 exons (coding pieces) and 2,351 amino acids—was cloned by a group of scientists from the San Francisco Bay-area biotechnology giant, Genentech.

Treatment for hemophilia has been revolutionized with each new bit of understanding about the disease. Until the early 1960s, hemophiliacs were given whole blood or fresh plasma to treat bleeding incidents. Neither blood product contains enough Factor VIII or Factor IX to stop serious bleeds, however, and most people with hemophilia before that time died by early adulthood.

A breakthrough came in 1965 when Dr. Judith Pool of Stanford Medical Center discovered that the sludgy gunk on the top of thawing plasma known as cryoprecipitate contained concentrated Factor VIII. As a result, hospitals could have enough Factor VIII on hand to make surgery possible for patients with hemophilia A and to stop life-threatening internal bleeding events when these patients showed up in the emergency room. Through the 1970s, researchers refined and expanded on Pool's work and were able to create freeze-dried powdered concentrates of Factor VIII and Factor IX that could be kept and used at home. Hemophiliacs could now enjoy a normal life filled with school, work, and hobbies. Unfortunately, all of these freeze-dried factor products are blood-based and are derived from vast quantities of pooled plasma. Indiana teenager Ryan White was a tragic example of how the blood products used to treat hemophilia could transmit HIV, and he was not alone. It is estimated that half of all hemophiliacs in the United States were infected with HIV during the late 1970s to the mid '80s.

In the wake of such infections, researchers through the 1980s and early '90s worked to develop methods to treat and purify these concentrated products to prevent viral contamination. At the same time, the identification of the genes encoding Factor VIII and Factor IX opened up the possibility that Factor VIII and Factor IX could be produced through recombinant DNA technologies. Producing these proteins in special hamster cells keeps them bacteria- and virus-free. In 1993, a recombinant form of Factor VIII became available, and in 1999 recombinant Factor IX was marketed as well.

Now, researchers are looking for ways to use gene therapy to allow a hemophiliacs' own body to start producing the coagulation factors they are missing. The work is ongoing, but some researchers think it is quite likely that hemophilia will be the first common genetic disease to be "cured" by the use of gene therapy.

On the other hand, almost 80 percent of the world's hemophiliacs can't afford replacement therapies and live in a world where hemophilia still takes its tragic and historic toll. Long before the genetics of the disorder was known, parents of affected boys knew the horror of the disease. They watched helplessly as their children died in childhood or early adulthood after suffering a minor injury that triggered uncontrolled bleeding. Sometimes the bleeding started spontaneously with no obvious impetus. Those hemophiliacs who did survive were crippled and deformed by repeated painful bleeds into joints such as knees, elbows, and ankles. Hemophiliacs can suffer prolonged bleeds in any part of the body, but bleeding into the brain, gastrointestinal tract, and neck are all life-threatening.

This was the reality in 1853 for Queen Victoria when her son Leopold was born with the disease. The confounding part for Victoria was that while other genetic diseases such as porphyria were known among the royal families in Europe, hemophilia had never before surfaced. Because Leopold had inherited his bleeding tendency on the X chromosome inherited from his mother, Queen Victoria was definitely a carrier of the disease even though her father wasn't a hemophiliac, and there is no evidence indicating that her mother was a carrier of the disease. One possible explanation—perhaps even the likely one—surrounds Victoria's father and his age. Edward, Duke of Kent, was fifty-two when Victoria was born. While much attention is paid to the age of a mother when she gives birth to children because of the likelihood of genetic conditions such as Down's Syndrome, older fathers produce sperm that has a higher frequency of mutations. Victoria's faulty Factor VIII gene may well have come from a spontaneous mutation in the X-chromosome donated by her father.

Irrespective of its origin, Victoria's daughters Beatrice and Alice brought hemophilia to the royal houses of Europe through marriage. Beatrice married Prince Henry of Hesse and bore two hemophiliac sons, a normal son, and a carrier daughter, Eugenie. Their daughter married Alfonso, king of Spain, thereby delivering the genetic disease to the Spanish royal family.

Alice married Louis, duke of Hesse, and bore six children: one hemophiliac son, two carrier daughters, and three noncarrier daughters. Their daughter Irene married Prince Henry of Prussia and brought the disease to Prussian royal family. It was Alice's daughter Alix whose marriage into the Romanov family that proved the most fateful: Alix married Nicholas II of Russia, and on November 26, 1894, became the Empress Alexandra Feodorovna.

By all accounts, this was a true love match. Nicholas defied his father's wishes that he marry a French princess to strengthen political alliances and married Alix instead. The couple had four daughters—Olga, Tatiana, Marie, and Anastasia—before their long-awaited heir, Alexei, was born in 1904. Unfortunately, Alexei was born with hemophilia.

Alexandra was devastated by her son's condition and sought a cure for the child's dangerous bleeding incidents. She thought she found such relief from the Siberian mystic Grigory Yefimovich Rasputin, who was renowned for his healing talents. In 1908, the imperial family summoned Rasputin to the palace during one of Alexei's bleeding episodes. Rasputin calmed the boy and staunched the bleeding. It remains a mystery how Rasputin was able to treat Alexei's bleeding; some have suggested it was a combination of hypnosis and medications. Nevertheless, he continued to minister to Alexei during uncontrolled bleeds and in the process won the favor of the empress.

Unfortunately, Alexandra had placed her trust in a wholly unholy man. The profligate Rasputin preached the only way to God was through sin . . . and lots of it. Rasputin, widely known as a heavy drinker, frequented prostitutes and maintained a number of mistresses. His scandalous personal life didn't sit well with most members of Russian society.

Nicholas and Alexandra's reign in Russia began with turmoil when a deadly riot broke out during the festivities associated with their 1896 coronation celebration. Throughout their reign the country was beset with strikes and discord. A disastrous war with Japan from 1904 to 1905 led to mutinies in the army and navy, and eventually forced Nicholas to make concessions and form the representative elected body called the Duma. In 1914, Nicholas involved Russia in World War I; in 1915 he left the country to personally lead his troops.

With Nicholas at war and Alexandra left to run internal Russian affairs, Rasputin seized power. Rasputin sacked competent ministers and filled the positions with cronies. With the country at war abroad and ill-managed at home, Russia was ripe for revolution. In 1916, conservatives murdered Rasputin in order to remove his influence over the empress and the country. It was too late, however, to save the monarchy.

On March 8, 1917, riots broke out in St. Petersburg, and Nicholas sent in the army to quell the violence. The government fell and Nicholas abdicated the throne to his brother Michael—who refused it—and so a provisional government was established. The family was confined to their palace and eventually moved to Siberia where they survived the Bolshevik revolution in October of 1917. The Bolsheviks moved the family to Yekaterinburg in November 1917, and in July of 1918 the family was executed.

Even if Rasputin hadn't been in the picture, it isn't clear whether Nicholas II might have been able to lead his country through the difficult times it faced. What is clear is that Rasputin helped to isolate Nicholas and his family so that they were blind to the public and political events around them. Had Alexei not been ill, the Romanovs might never have made Rasputin's acquaintance. So, in some small way, one family's struggle with hemophilia altered history for the entire world.

Even though it's a true flight of fancy, one can't help but wonder if recombinant Factor VIII and Factor IX had been available a century earlier, might the history books have had the same chapters on the Russian revolution, World War II, and the Cold War?

The Fragile X

Since the nineteenth century, physicians have noted that mental retardation strikes boys with greater frequency than it does girls. For almost two centuries, in the institutions dedicated to the care of the mentally retarded, boys have outnumbered girls roughly five to four. Among those boys, a great many have large, protruding ears and long faces.

The fact that boys are more likely to suffer from mental retardation isn't surprising when you take into account that no less than 130 X-linked syndromes cause mental retardation. Far and away the most common cause of that X-linked overrepresentation of boys suffering from inherited mental retardation is fragile X syndrome.

While it may be relatively common—fragile X syndrome affects one out of every four thousand males and one out of every eight thousand females—everything from the genetic defect causing the disorder to its role in causing disease has offered new scientific surprises.

First known as Martin-Bell syndrome (an X-linked condition), most boys with fragile X syndrome are moderately to severely retarded. Once they hit adulthood the facial traits—a long face with a big jaw, prominent ears, and a high arched palate—become more pronounced. Tall as children, males with fragile X syndrome are often short as adults. They tend to have highly flexible joints, flat feet, and large testicles. In addition, their voices can be high pitched, and they repeat themselves often. Children with fragile X often suffer from behavioral and emotional disturbances as well.

Females who develop fragile X syndrome have a much wider range of symptoms or phenotypes. Some of these girls exhibit nothing more than mild learning disabilities; in others the disease manifests as severe retardation with most of the physical characteristics typical among the affected males.

It wasn't until 1969 that the disease received the name "fragile X syndrome." Herbert Lubs, an American geneticist, discovered that the boys suffering from the condition had distinctive X chromosomes: when placed under a microscope, the tip of the long arm of the X chromosome appears to be broken, almost as if it were dangling from a thread. Lubs had discovered a critical feature of the disease and a means of testing for it. Testing, however, didn't become widespread until the late 1970s when Grant Sutherland, from the Adelaide Children's Hospital in South Australia, developed a procedure for stabilizing the chromosome for testing.

With a means of identifying the condition, researchers discovered that fragile X syndrome was the most common form of inherited mental retardation. They also found the condition had some unusual features. While the disease is X-linked, unlike hemophilia, fragile X syndrome displays dominant inheritance. In other words, females need to inherit only one copy of the mutated gene, not two, in order to be affected. However, the disease is far more mild in girls than boys and is less "penetrant"—80 percent of the boys who inherit a mutated fragile X gene will develop mental retardation but only 30 percent of the girls will. This finding is most likely explained by the fact that each cell in a girls body only relies on one X chromosome; and if enough cells rely on the normal chromosome, the disease is less severe.

By the 1980s, fragile X syndrome was proving a puzzlement. Stephanie Sherman from Emory University noticed something unusual about the way the condition was inherited: fragile X syndrome didn't manifest itself in future generations with the same probability as other X-linked diseases—a phenomenon that came to be known as the "Sherman paradox."

Males who carried the gene but didn't have the disease—known as transmitting males—passed the gene *and* fragile X syndrome to 40 percent of their grandsons and 50 percent of their great-grandsons. Daughters of transmitting males NEVER developed fragile X syndrome even though their sons often did. This pattern of inheritance violates one of the tenets of X-linked inheritance: the percentage of males developing the condition as a result of inheriting the gene *should* remain stable.

This unusual pattern of inheritance wasn't resolved until 1991 when the gene for fragile X syndrome FMR1 was isolated by an international team of researchers. The FRM1 gene contains 17 exons and is 38,000 base pairs long. It was Australia's Grant Sutherland who discovered a critical clue: a series of three nucleotides, cytosine, guanine, guanine (CGG), repeated anywhere from 6 to more than 1,000 times in the area just before the FMR1 gene. Much like highway signs warning of exits and on ramps, the regions directly adjacent to genes tend to serve as signposts telling the cellular transcription machinery to take note that a gene is in the vicinity. These regions are called promoters.

The repeated region has between 6 and 50 copies in normal individuals. People who have fragile X syndrome carry between 230 and 1,000 or more copies of the repeat and are referred to as having the "full mutation." It's those individuals with repeat numbers in between 50 and 230 that prove most interesting. This intermediate number of repeats is referred to as the "premutation," and it can be found among the transmitting males and their daughters. For reasons that still aren't clear, when a daughter with a premutation passes the gene on to the next generation, the number of repeats increases substantially and may even hit the magic number of 230 repeats, thus causing the full mutation. The number of repeats never increases when a male passes the gene onto his children. Each time a female passes the premutation on to the next generation, the number of repeats increases until it eventually expands to the full mutation level.

While this inheritance pattern sounds confusing, the ability for triplet repeats to expand elegantly explains the Sherman paradox. Males with the premutation pass an X chromosome only to their daughters—their sons receive a Y chromosome. The triplet repeat can expand to a full mutation when the daughters of transmitting males pass an X chromosome with the defective FMR1 gene on to their sons, causing fragile X syndrome. However, if the number of repeats expands only to a higher number within the range of the premutation when the daughter passes her X chromosome to the next generation, both her sons and daughters will be unaffected. When the transmitting male's granddaughter subsequently has children, the likelihood that the premutation will expand to the full mutation increases and her sons (the original transmitting male's great-grandsons) are more likely to suffer from fragile X syndrome. That's why a transmitting male's great-grandsons are more likely to have fragile X syndrome than are his grandsons.

The discovery of the FMR1 gene was the first instance where a triplet repeat had proven the cause of a genetic disease. Since then, researchers have shown that other neurological conditions such as Huntington's disease, myotonic dystrophy, and spinocerebellar ataxia type 1 are caused by repeated nucleotide triplets. How these triplet repeats cause disease, however, differs from gene to gene. In the case of fragile X syndrome, the lengthy repeat of CGG nucleotides causes the cells to chemically alter the promoter region of the FMR1 gene and block efforts to transcribe the gene into RNA thereby silencing the gene.

Researchers knew the full mutation results in silencing the gene, thereby preventing the cell from producing the protein known as fragile X mental retardation protein (FMRP). But in 1991, they didn't know how the absence of a single protein could lead to mental retardation and behavior problems. In the fourteen years since the discovery of the FMR1 gene, much has been learned about fragile X syndrome even though the mechanism by which the cell distinguishes between the full mutation and the premutation remains a mystery.

Since its discovery, the premutation in fragile X syndrome has been viewed as merely a benign harbinger of disease for future generations. That premise was shattered in 2003 when Paul and Randi Hagerman from the University of California–Davis reported that adult carriers of the premutation can develop a movement and tremor disorder called fragile X tremor/ataxia syndrome (FXTAS) as they age. Carriers of the premutation also appear to be susceptible to emotional problems such as anxiety, obsessional thinking, and depression.

In patients with FXTAS, the FMRP protein appears to bind to itself and form pockets of protein called inclusion bodies in the nucleus of brain cells. The Hagermans found these inclusion bodies to be unlike any others found in diseases like Parkinson's disease. As a result, they argue that the premutation in patients with FXTAS produces some gain in function for the FMRP protein that proves toxic to brain cells. What remains to be seen is how the same gene can use different mechanisms to give rise to not only the most common inherited form of cognitive impairment in children but also a neurological disorder of aging.

Identifying the function of normal FMRP protein will be critical that understanding. In 1996, Stephen Warren and his group at Emory University suggested a role for FMRP in controlling the production of protein in brain cells. They discovered that the protein shuttles between the nucleus and the cytoplasm of a neuron ferrying the templates for protein production—messenger RNAs. Because of its role shepherding messenger RNA, researchers have begun to speculate that FMRP plays a critical role in learning.

Rather than neat little balls, nerve cells are elongated cells that form synapses between projections sticking out from the nucleus-containing cell body. Projections from the nerve cell which receive signals from the synapse are called dendrites while axons are the projections that move the signal along to the next synapse. Nerve cells have protein-making machinery at the very tip of the dendrite, presumably to allow the signal from the axon across the synapse to propagate quickly through the nerve cell.

FMRP shuttles messenger RNA from the nucleus to the tip of the dendrite and associates with the protein making machinery. It also has a penchant for associating with tiny inhibitory bits of RNA that bung up the works and stop the cellular machinery from producing protein. Warren suggests FMRP may actually inhibit the production of protein from the messenger RNA it ferries out of the nucleus.

While not intuitively obvious why this would benefit a nerve cell, it makes some sense when you consider what a nerve cell must do to permit learning: it must receive a signal, send the signal along the neural pathway, and be ready to receive the next signal rapidly. FMRP may be a sort of reset button that allows the neuron to prepare itself for the next signal by halting the production of protein necessary to propagate a signal.

Those inhibitory RNAs may also explain how the cells come to modify the promoter region of the full mutation FMR1 gene. While cells have intricate systems to maintain the double-stranded nature of DNA, they have just as many systems in place to degrade double-stranded RNA. Nature, it seems, abhors double-stranded RNA. Karen Usdin and colleagues at the National Institute of Diabetes, Digestive, and Kidney Diseases in Bethesda, Maryland, discovered that when the promoter region of the full mutation FMR1 gene isn't chemically altered, it forms a long RNA that loops back on itself to form double-stranded RNA.

Warren proposes that during the earliest stages of development the full mutation allele of FMR1 hasn't been modified and is available to be transcribed. When it is, it will form a double-stranded RNA, which then becomes a target for cellular proteins that chop it up. The resulting little bits of RNA can bind to the promoter region and serve as a signal to modify and ultimately silence FMR1.

While much about the role of FMRP has yet to be proven, it's clear that fragile X still has more surprises in store for us to learn.

The Werewolf Gene

Step right up! See the amazing Mexican werewolf! Marvel at his primitive physiognomy!

It's sad but true. People who developed excessive hairiness (hypertrichosis) often ended up listening to a circus barker day in and day out describe them as "dog man" or "ape man." Most of the time, these people could blame their condition on a hormonal imbalance. A few could blame their genes.

Surprisingly, one large five-generation Mexican family can lay fault squarely on mom for the sudden expression of this long dormant gene that causes their hairiness (a form of congenital generalized hypertrichosis—CGH). In 1995, Pragna Patel, of Baylor College of Medicine in Houston, identified a region on the far end of the long arm of the X chromosome that displays an X-linked dominant inheritance in this family.

While previous studies had indicated that the gene displayed dominant inheritance, a close look at the transmission of the gene in this Mexican family indicated the trait was only passed on through the generations by females. The trait isn't expressed identically in males and females, however. Males are usually hairier, and its appearance on the face and upper body is complete. Females, on the other hand, usually display a patchier pattern; probably because each cell uses only one X chromosome, some of the cells will use the X chromosome with the gene for hairiness while others will use the chromosome with the normal gene.

A complex interaction of genetic and endocrine factors control human hair growth. Most forms of excessive hairiness are associated with hormone imbalances involving body sites under male hormone, or androgen, control, such as the face, and are known as hirsuitism. On the other hand, the excessive hairiness associated with hypertrichosis can involve any area of the body. Patel believes finding the exact gene for the X-linked dominant form of CGH should help researchers to understand the molecular processes that control hair growth and hair distribution.

Nonhuman primates have considerably more hair than we do because genes for that level of hair coverage have undergone significant structural or regulatory changes during the course of evolution from chimpanzees to humans. Humans have much less hair and rarely experience the total body hairiness characteristic of their closest living relative. For humans overall hairiness represents the reappearance of an ancestral characteristic—a genetic atavism.

Once the source of embarrassment for evolutionary biologists because they flew in the face of the concept that evolution progresses toward ever more complex and superior organisms, atavisms now serve as proof that a phenomenal amount of genetic information is retained after a physical structure disappears from a species. For example, all modern horses evolved from an ancient three-toed horse to have one functional toe ending in the modern hoof. Brian K. Hall, a biologist at Dalhousie University in Halifax, points out that, occasionally, a modern multitoed horse reappears because at some point in embryonic development, a gene that normally turns off the growth of the additional toes in the modern horse fails to do so. As a result, the toes continue to grow as they had on the ancestral creature. In other words, the ability to grow the extra toes or to have total body hairiness wasn't lost during evolution, it was merely silenced.

Genetic atavisms arise in other animals as well. Roughly one out of every five thousand whales develop hindlimbs. That happens not because a mutation suddenly creates the ability to develop the limb, but because a mutation prolongs the whale embryo's limb bud development at a time when it is normally shut down.

In the case of excessive human body hair, it remains to be seen exactly which gene unleashed this ancient pattern of hair growth and how that gene functions. Based on the fact that people with CGH also develop other abnormalities, Patel suspects the gene may encode a growth factor or its receptor, adhesion molecules, or an enzyme involved in the metabolism of connective tissue. Whatever gene is involved, it seems likely we'll soon learn something more about hair growth and what makes us different from our nearest evolutionary ancestors.

The Cue Ball Gene

Their forefathers had patent medicine cure-alls like Vegetable Balm and other snake oil to slather onto their bald pates in the hopes of reviving the luxuriant youthful locks that had fallen out. Today's balding men have minoxidil, hair transplants, and

Propecia® as a means to battle the dreaded comb-over. What they all may have in common is a mother whose father shared their shiny-headed fate.

Male-pattern baldness, the bane of an estimated 35 million American, mostly Caucasian, men begins with thinning hair at the temples. Then the thinning takes root at the back of the head. By the time back meets front in a smooth shiny scalp, thousands of dollars may have been spent trying to hang on to each and every precious strand. When the hair loss begins and how much is ultimately lost vary from person to person. The only sure thing about male-pattern baldness, or androgenetic alopecia, is that hair loss depends on the presence of androgens—male hormones such as testosterone and dihydrotestosterone.

Susceptibility to male-pattern baldness is subject to the influences of a number of different genes, but *which* genes play a role has been a mystery. A group of scientists in Germany believe they have uncovered one of those genes. This one lies on the X chromosome. Because every man inherits his X chromosome from his mother, at least some baldies can blame mom.

German researchers, headed by Markus Nöthen, of Bonn University, and Roland Kruse, Düsseldorf University, began a genome-wide scan looking for genes associated with male-pattern baldness. The researchers looked at 198 males who suffered from early onset baldness, 188 control men, and 157 men with full heads of hair.

What they discovered was the gene that encoded the androgen receptor. Hormones like testosterone, estrogen, and progesterone have their effects by binding to receptor proteins on the surface of target cells such as testis cells and breast cells, and setting off a cascade of chemical reactions. If androgen exposure is a cardinal prerequisite to male-pattern baldness, it stands to reason that a gene involved in androgen signaling would be a good candidate for a baldness gene.

The androgen receptor is a huge gene comprised of eight exons—the sections of DNA that are actually read by the cellular machinery to produce proteins. The German researchers found a curious situation in the first androgen receptor exon of the men who'd begun losing their hair early in life: a set of three nucleotides coding for the amino acid glycine repeated as many as twenty-three times.

However, when this set of nucleotides was repeated twenty-four times, the men were much less likely to suffer from male-pattern baldness. Even more curious is the fact that the shorter alleles in this region have been associated, albeit with some controversy, with prostate cancer. Still, some researchers have suggested a relationship between prostate cancer and male-pattern baldness. In addition, one study has shown a relationship between the longer glycine-bearing alleles and endometrial cancers.

Finding the relationship between the androgen receptor allele and male-pattern baldness does nothing to explain how a smaller number of glycine repeats in the androgen receptor could lead to male-pattern baldness. The researchers speculate that the shorter allele either produces more androgen receptors in the scalp, or it causes the androgen receptor to last longer. Either situation would result in getting more bang for your buck from androgen hormones. Unfortunately, a bigger dose of androgens in the scalp means fewer hairs on that same scalp.

The German researchers estimate that the shorter glycine-bearing allele of the androgen receptor is almost 50 percent responsible for male-pattern baldness. That would leave the rest of the blame on other, as yet unidentified, genes. Those genes, however, are unlikely to be found on the X chromosome. Given the German researchers discovery, you would expect to see a man's hair-loss pattern resemble that of his maternal grandfather. But most balding men's heads resemble those of their fathers and brothers. As a result, other genes associated with male-pattern baldness will most likely lie on chromosomes that allow the susceptibility to baldness to be inherited from dad.

So it's still possible that an equal helping of balding blame can be served to dad, but mom's role appears to be the first to be discovered.

chapter Four

Leaving an Imprint

For all the talk of disease-causing genes, the fact that those genes don't tell the whole story is sometimes lost in the excitement of unraveling a new medical mystery. How those genes turn on and off, and what environmental influences make it more likely that they will become active or remain quiescent, involves far more than the "simple" sequence so recently provided by the Human Genome Project.

As a cell moves from early embryo to the specific tissues that make up the body such as skin, liver, heart, and brain cells, gene activity waxes and wanes for different reasons during the various stages of development. It's a subtle yet tightly controlled symphony where each gene enters the movement at a precise time to create a healthy human being. In order to achieve such subtle control, the cell marks certain genes meant to be silent. How these cell marks affect gene expression is the purview of the emerging field of epigenetics, quite literally "on genes."

It's important to remember that chromosomes aren't naked strands of DNA; they are amalgamations of DNA and structural proteins known as histones. Any modification to either the DNA or the histone will affect the ultimate expression of a gene in a cell. These exceedingly subtle modifications adorn individual nucleotides of the DNA strands and single amino acids in histones. DNA is dotted with alterations called methyl groups: a carbon atom surrounded by three hydrogen atoms attached to individual DNA nucleotides. Far from mere decoration, this DNA methylation controls gene expression. In order for a cell to use the information stored in DNA to produce a useful protein, the cellular machinery must recognize the gene and begin to transcribe it into RNA, which is then used as the template for another set of cellular machinery to produce protein. The DNA sequences flanking a gene are critical for getting the cellular machinery to take notice that a gene is in the area. Marking those regions with methyl groups effectively gums up the works, thus preventing the cellular machinery from expressing the gene.

Methylated DNA can directly shut off gene expression by physically blocking the transcriptional machinery. It's more likely that the methyl modifications attract proteins that alter the histone so it causes that section of the chromosome to condense and prevents gene expression.

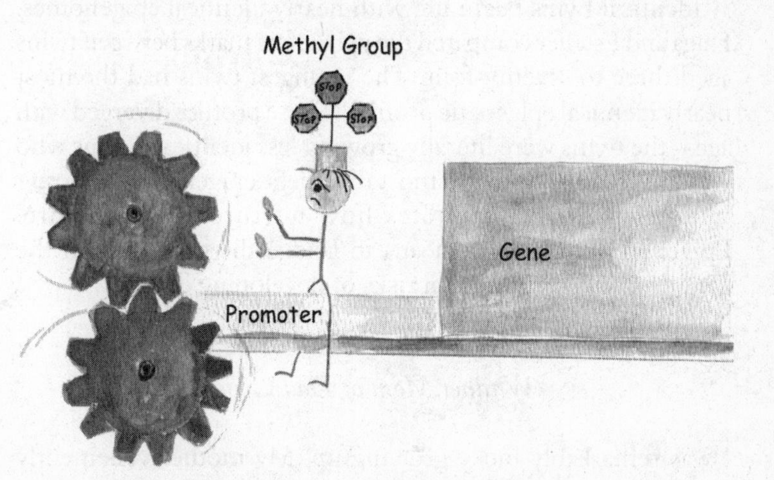

The area in front of a gene, the promoter, contains DNA sequences that alert the cellular machinery that a gene is in the vicinity. Tagging certain nucleotides with a methyl group gums up the works for the cellular machinery trying to transcribe the gene. As a result, the gene falls silent.

Scattered across the genome, these epigenetic alterations make sure that genes needed for a cell meant to be a liver cell aren't expressed in skin cells. In addition, they play critical roles in cancer. A gene that suppresses tumor growth needs to be active. If it is inappropriately methylated, the result can be cancer. The opposite holds true for genes that spur cancer growth. These genes can become active if the DNA surrounding them suddenly lose their epigenetic marks.

Remarkably, the exact placement of epigenetic marks on the genome is highly personal, and these marks change over time in response to such environmental exposures as cigarette smoke, certain foods, and pollutants, among other things. In 2005 Mario F. Fraga and Manel Esteller of the Spanish National Cancer Center in Madrid demonstrated the degree to which the epigenome changes throughout our lives by analyzing epigenetic alterations in identical twins. Fraga and Esteller focused their study on DNA methylation and histone acetylation: DNA methylation silences a gene; histone acetylation reverses gene silencing.

Identical twins begin life with nearly identical epigenomes. Fraga and Esteller compared the epigenetic marks between twins aged three to seventy-four. The youngest twins had the most nearly identical epigenetic profiles. These profiles diverged with age—the twins were literally growing less identical. Twins who were raised apart had the most divergent epigenetic profiles.

This work demonstrates how nurture affects nature. Epigenetics provides a means to bridge the gap between the genetic and environmental risks of developing diseases.

Whither Mom or Dad Gene

I was remarkably lucky growing up. My mother vehemently believed women were just as capable as men and that they could pursue any career path men did and succeed. At least I think this is what she believed. It is quite possible she was merely trying to serve as a foil to my father who periodically disparaged the talents of "women engineers" while encouraging me to excel in math and science. It was a little confusing, and my father to this day denies he ever complained about the talents of females in science. However, no harm, no foul, and I only grew in my conviction that men and women are indeed created equal. It's a value that I fully intend to transmit to my daughter even though feminism may now be grossly passé.

Unfortunately, I've hit upon a little snag in my belief system: Mother Nature. While I believe it doesn't matter what a male or female contributes to any project, when creating a new human, Mother Nature sometimes cares very much whether a gene comes from mom or dad. It's a curious situation that flies in the face of everything Mendel taught us with his peas.

When Mendel completed cross after cross of wrinkled peas and smooth peas and purple-flowered peas and white-flowered peas, he proved that it didn't matter whether the trait came from pollen or ova—the ratio of the phenotypes would always be the same. This genetic dogma has served as a mooring for genetic studies and analysis since his work was rediscovered.

But it seems like every time we humans get a safe mooring in our understanding of genetics, the complexity of the biological system reveals itself and leaves the scientific community adrift looking for explanations. This time, the explanation is extraordinary: at the earliest stages of development, some maternal genes are turned off to make room for the paternal genes and vice versa. The reason for this may just be a battle royal over whose genes speak in and hold influence over the next generation.

The process at work is called genomic imprinting. It's a phenomenon where certain genes are expressed depending on whether they have been inherited on the maternal or the paternal chromosome. In humans, defects in genomic imprinting reveal themselves in a unique pair of congenital anomaly syndromes: Angelman syndrome and Prader–Willi syndrome.

The two disorders are strikingly different. Children suffering from Angelman syndrome display severe mental retardation and seizures. Many children with Angelman syndrome never walk, or if they do they walk with a wide-based and stiff-legged (ataxic) gait. Despite mental impairment, children with Angelman syndrome have an enormously affectionate nature and frequently burst into laughter. Even so, these children often laugh at inappropriate situations, like when they are in pain. Children with Prader-Willi syndrome suffer only mild to moderate retardation. The have short stature with small feet and small hands and, at birth, very poor muscle tone. Most of the babies born with Prader-Willi syndrome are described as "floppy." These children are also prone to temper outbursts and stubbornness. One of the salient features of a child with Prader-Willi syndrome is an insatiable appetite, which generally leads to obesity.

Best estimates indicate both these conditions strike approximately one out of every fifteen thousand individuals, and both males and females and all ethnic groups are affected equally. In addition, 70 percent of cases can be linked to nearly indistinguishable defects on chromosome 15. The two syndromes are most often caused when a chunk of DNA (called the critical region) on chromosome 15 is deleted. Whether a child develops Angelman syndrome or Prader-Willi syndrome depends entirely on whether that chromosome with the deleted portion came from mom or dad.

By the late 1980s, researchers had discovered that when the chromosome with the deletion including the critical region is inherited from mom, the child will be born with Angelman syndrome. When the same chromosome defect is inherited from dad, the child will develop Prader-Willi syndrome. How a newly formed embryo distinguished between the maternal and paternal chromosomes was a mystery.

It wasn't going to remain a mystery for long. In 1987, Wolf Reik and colleagues from Cambridge University discovered that a DNA methylation identified whether the chromosome came from mom or dad. Reik and his colleagues, studying mouse embryos, discovered that DNA methylation determined whether a chromosome comes from mom or dad. Certain genes on the maternal chromosomes were methylated, whereas the same genes on the paternal chromosomes remained unmethylated. Likewise, the methylated genes on the paternal chromosomes remained unmethylated on the maternally derived chromosomes. In other words, DNA methylation proved a means of marking and imprinting the origin of a chromosome.

In 1992, Robert Nicholls, now at the University of Pennsylvania, and colleagues at the University of Florida College of Medicine discovered that, like the mouse, differential DNA methylation was at work on human chromosome 15 as well. When a person inherits a maternal chromosome 15 with a deletion in the critical region, genes normally active on the maternal chromosome 15 are missing. Their counterparts on the paternal chromosome have been rendered silent by DNA methylation causing Angelman syndrome. Inheriting a paternal chromosome 15 containing a deletion in the critical region silences important genes because the maternal counterparts are inactivated by DNA methylation, resulting in Prader-Willi syndrome.

While these deletions account for the bulk of both Angelman and Prader-Willi syndrome cases, other situations can cause the same diseases. Sometimes a child with one of these conditions inherits two chromosome 15s from one parent. This happens in one of two ways: A child may inherit two maternal chromosomes when an egg is carrying chromosome 15 in duplicate rather than as a singlet. Upon fertilization, the zygote then has three copies of that particular chromosome, a condition known as a trisomy. Trisomy 15 is usually fatal to a fetus. Scientists believe that an overloaded zygote ejects one of the extra chromosomes, and by chance one-third of the time the cell keeps only the maternal chromosomes. A double dose of maternal chromosome 15 results in Prader-Willi syndrome because essential genes, which are usually expressed from the paternal chromosome, are silenced on both of the maternal chromosomes. Conversely, when there is a double dose of paternal chromosome 15—usually as a result of a chromosome duplication following conception, an exceedingly rare event—the child develops Angelman syndrome.

It appears, in fact, that several imprinted genes are involved in Prader-Willi syndrome. Angelman syndrome is a different story. In 1997, two teams of researchers led by Joseph Wagstaff of Harvard Medical School and Arthur Beaudet of Baylor College of Medicine, identified a gene known as E6-AP ubiquitin protein ligase (UBE3A) as the culprit in Angelman. The researchers speculate that absence of the gene activity may cause proteins to build up in the brain.

While scientists are making remarkable progress in figuring how imprinting causes conditions like Prader-Willi and Angelman, the fact that imprinting exists at all is difficult to explain. The elegant thing about having pairs of chromosomes is that in most cases if a gene on one chromosome is faulty, a normal gene on the other chromosome of the pair serves as a backup. In the case of genomic imprinting, a nearly inconceivable thing occurs: the cells in the body silence one copy of the gene, leaving only one active copy. If something goes wrong with that copy, then an affected subject better be able to do without that particular gene product.

Genomic imprinting is not a logical evolutionary development. On the surface, it's hard to see how there could be an evolutionary advantage in throwing away one's safety net. Yet, this process appears to be a relatively recent development in the animal kingdom—imprinted genes are found only in mammals and marsupials. Therefore, the phenomenon must deliver some sort of advantage.

One of the most intriguing explanations was posited in the early 1990s by David Haig of Harvard University and Tom Moore of Cambridge University. Their hypothesis maintains that certain genes from the mother and those from the father are at odds with each other. Basically, their model holds that as far as dad's genes are concerned, bigger is better; and mom, she just wants to make that sure the current pregnancy isn't her last one.

This genetic conflict model gains credence because the phenomenon of imprinting exists only in animals that form a placenta—an organ that serves as the interface between a fetus and a mother's womb—and are dependent on mom for prenatal nutrition. (Actually, this isn't *entirely* true. Imprinting evolved independently among seed plants. These plants clearly don't have a placenta, but the endosperm plays a similar role.)

After fertilization, the father needn't expend biological resources to ensure the growth and survival of the fetus. All of the biological investment comes from the mother. As such, the genetic conflict model predicts dad's genes will spur the growth of the fetus because a bigger, better-fed offspring is more likely to survive to reproductive adulthood and pass on dad's genes. Mom's genes, on the other hand, will tend to put the breaks on growth because mom not only needs to ensure the survival of the fetus she is carrying, but she also needs to survive to nurture and bear more children.

Even so, establishing and maintaining imprinting requires more than an unequal contribution to fetal survival. The animals must also be promiscuous. The genetic conflict arises when mom has children with a number of different fathers. By turning off genes promoting growth, mom can maximize the survival of all of her offspring. Dad, on the other hand, maintains no interest in the half-siblings of his offspring. As a result, his genes will promote the growth of his offspring at the expense of offspring that aren't his.

If this genetic battle is a good explanation for imprinting, then genes that spur growth would be prevented from being active or imprinted on the chromosome inherited from the mother, while the genes that limit growth would be imprinted on the paternal chromosome. One gene pair fits this profile exactly: the genes encoding insulin-like growth factor 2 (IGF2) and the insulin-like growth factor receptor (IGF2r).

In a normal placental embryo, the maternal copy of IGF2 would be turned off through imprinting because IGF2 encourages growth, and the paternal copy of IGF2r would be inactive because its role is to bind to IGF2 and trigger its degradation. This scenario is, in fact, exactly what happens. What's more, mice with an inactive paternal copy of IGF2 suffer from fetal growth problems called intrauterine growth restriction. Mice with an inactive maternal copy of IGF2r grow to be much larger then normal.

Another human genetic disease lends support to this battle of the sexes theory. Beckwith-Wiedemann syndrome causes fetuses to be large for gestational age. Children born with the disease have large tongues and have an abdominal wall defect because their internal organs grow so large. In addition, children with Beckwith-Wiedemann syndrome often develop a type of kidney cancer known as Wilms tumor. These children experience imprinting problems on chromosome 11 . . . where the IGF2 gene resides. Most often, the maternal IGF2 gene escapes imprinting in children with Beckwith-Wiedemann syndrome and is therefore active. Sometimes, a child with Beckwith-Wiedemann syndrome will have two copies of the paternal chromosome 11. Either way, the developing fetus gets a double dose of IGF2 and grows quite large.

IGF2 and IGF2r aren't the only genes identified in the mouse supporting the genetic conflict theory. PEG1, PEG3, and insulin are all expressed only on the paternal chromosome. When mice are bred that lack the paternal contribution of these genes, the fetuses suffer intrauterine growth restriction. Similarly, mice bred without the maternally expressed gene H19 produce unusually large fetuses.

There are, however, a few glitches with this popular explanation for imprinting. The first is that not all imprinted genes play crucial roles in the development of a fetus; for example, UBE3A, the gene associated with Angelman syndrome, isn't vital for fetal growth. There are plenty of genes regulating fetal growth that aren't imprinted, such as insulin-like growth factor 1 (IGF1). It is possible that some genes may have been accidentally imprinted when a nearby gene that does regulate growth was imprinted. In addition, the conflict theory doesn't demand that all growth-stimulating genes bear imprints.

Perhaps the most difficult aspect to prove about the genetic conflict theory has been the requirement that any members of imprinted species have multiple mates over their lifetimes. Mating for life is exceedingly rare in mammals. But it stands to reason that if imprinting were a necessary mechanism of dealing with multiple mates, faithful species then have no need for it, and their genes shouldn't be imprinted.

Because not all species of mice play the field, Shirley Tilghman's group at Princeton University decided to test this aspect of the theory by mating a faithful species of mice with a philandering species of mice to see whether there were any differences in imprinting. When they mated the philandering father with the monogamous mother, the placenta was six times larger than when the monogamous father was mated to the philandering mother. On the surface, this result bolsters the genetic conflict theory. However, when the researchers looked a little closer and examined which genes undergo imprinting in these two sets of mice, they got a surprise. The offspring of these two different matings imprinted the same genes.

Tilghman's result isn't a crucial refutation of the theory. First, it may be that not enough evolutionary time has elapsed since the point when the monogamous species took up its lifestyle. Second, the monogamous mice may not be truly monogamous, and that low levels of polygamy may be enough keep imprinting in place. In addition, mathematical models indicate that the periodic deaths of males of a monogamous species are also sufficient to keep imprinting in place. Walter Mills of the University of Cambridge and Tom Moore, now at University College Cork, discovered that a swinging lifestyle among females has its greatest effect on imprinting at low levels of promiscuity rather than at high rates.

Whatever the impetus for placental animals to develop an imprinting strategy, genomic imprinting is a critical component of mammalian DNA well past fetal development. When a normally imprinted gene such as IGF2 turns itself back on by escaping imprinting, the cell, which now has two active copies of IGF2, becomes cancerous. Roughly one hundred genes scattered about the genome are flying solo without any safety net. Their genetic activity depends on whether they came from mom or dad. In this case, no amount of feminist theory can change this basic biology. Nevertheless, I still intend to make sure that my daughter knows she can be a doctor, lawyer, construction worker, baker, well anything she wants to be . . . including an engineer.

The Calico Cat Gene

Women who seek to be equal to men lack ambition.

—A popular bumper sticker

Cats exist as proof that not everything in nature has a function.

—Unknown

To say that women are just more complex than men isn't a sweeping sexist generalization—it's simply a biological observation. Just look at the chromosomes that dictate male or female development: the X and the Y. Not to knock the Y chromosome, but throughout evolution the chromosome has shrunk to the point it houses only about a hundred genes, albeit vital ones if you happen to be a guy. The X chromosome, on the other hand, hosts over a thousand genes that prove crucial to both men and women.

Those genes are remarkably useful. They do things like clot blood, facilitate seeing colors, and mediating blood pressure, not to mention playing some role in deciding which males get to keep their hair. Because they carry an X and a Y chromosome, males get a single dose of the genes available on the X chromosome. Women have two X chromosomes and theoretically get a double shot of those genes. Therein lies the complexity. Nature doesn't take kindly to uneven doses of chromosomes. Embryos possessing merely a single copy of any other chromosome in the genome quite simply die.

In an attempt to keep things equal between the sexes, Nature shuts down one X chromosome in every cell of a woman's body. However, which X chromosome is shut down, the maternal one or the paternal one, is, for the most part, a crapshoot. As a result, some cells in a woman's body express the X chromosome genes she inherited from mom, and other cells express those she inherited from dad. As far as the X chromosome is concerned, every woman is a unique mosaic of genetic expression.

The characteristic black and orange patches of a calico cat develop because the gene for orange color resides on the X chromosome. One allele modifies black pigment to produce an orange color; the other allele produces only black color. Early in development, cells choose to inactivate one X chromosome or the other at random. When the cell inactivates the chromosome containing an orange allele, a black patch develops; inactivating the chromosome containing the black allele produces an orange patch. The calico cat is a visible genetic mosaic.

The calico or tortoiseshell cat is a perfect illustration of this fact. It's a simple truth that when you see a calico or tortie feline, it's almost always a female. The characteristic mottled coat pattern arises because the gene encoding the orange color resides on the X chromosome. One version, or allele, of the gene modifies black pigment to produce an orange color. The other allele produces only a black color. A calico or tortoiseshell cat has inherited an orange allele from one parent and a black allele from the other. Early in development, each cell of the embryo randomly shuts down one or the other X chromosome. The result is clumps of cells that express the orange coat color and clumps of cells that express the black coat color of the familiar calico pattern. To be entirely correct, there are, in fact, a few male calico cats. Their existence is explained because they carry two X chromosomes and one Y chromosome. Each cell in these unusual male cats shuts down one of its X chromosomes just like female felines.

The fact that different cells in a woman's body rely on genes coming from the maternal X chromosome while others utilize those from the paternal X chromosome provides a significant level of complexity in determining how an X-linked disease may or may not affect a woman. What's more, far more genes on the supposedly inactive X chromosome somehow manage to get expressed anyway. That finding is only the latest surprise in what has been a series of unexpected developments surrounding the X chromosome and its inactivation.

In 1948, armed with a $400 research grant, Canadian scientist Murray L. Barr and graduate student Ewart G. Bertram were trying to understand the impact that fatigue had on the brain by studying exhausted felines. The University of Western Ontario researchers were analyzing the changes in feline nerve cells during heightened nerve activity when they discovered some odd-looking clumps in the nerve cell nuclei they were studying. Barr and Bertram weren't sure what these clumps were, other than amalgamations of protein and DNA. But they knew one thing about these strange masses in the brains of cats: they found them only in the brains of female cats, never the males. They dubbed these blobs Barr bodies.

While the phenomenon had a name, Barr and Bertram still didn't know what these Barr bodies were. In 1959, Susumo Ohno, a geneticist and biochemist at the University of California, Los Angeles, answered that question when he discovered the Barr bodies were very condensed X chromosomes. That same year, mice with a single X chromosome were bred and shown to be viable fertile females. As a result, the idea that females needed only one X chromosome began to take root.

The preponderance of the evidence led Mary Lyon of Britain's Medical Research Council's Mammalian Genetics Unit in 1960 to propose that in every female cell one of the X chromosomes was inactivated and became a Barr body. It was a revolutionary proposal, but it so elegantly explained the Barr body data and Lyon's own research on mouse coat colors that skeptics were quickly silenced. At the time, X inactivation was referred to as Lyonization, a term some scientists continue to use to this day.

Lyon noted that shortly after she proposed her theory of X chromosome inactivation it became clear that one X chromosome couldn't be silenced entirely. There are, after all, some genes with counterparts, or homologues, on the Y chromosome. To keep strict dosage control in place, those homologues on the Y chromosome needed to be active on the "inactive" X chromosome as well. Thus started the search for how the X chromosome becomes inactive and which genes escape this inactivation.

Only mammals and marsupials employ the phenomenon of X inactivation to manage the dosage of genes. Other animals use various group-specific means to maintain dosage compensation. In general, both the maternal and paternal X chromosome are eligible for inactivation. Marsupials, however, preferentially inactivate the paternal chromosome, and some mouse cells do the same thing.

While Lyon proposed her theory in 1960, it was decades before researchers pinpointed how the cell actually inactivated one of its chromosomes. In 1974, Eeva Therman and Klaus Patau from the University of Wisconsin Medical College proposed the idea that inactive X chromosome condensation took place around an X inactivation center on the long arm of the X chromosome near the centromere. Part of their reasoning came from the fact that when an X chromosome carried two copies of the presumed X inactivation center, they formed Barr bodies with two "waists." In addition, a mouse locus for X inactivation had already been described.

In 1991, Huntington Willard and colleagues isolated a gene, dubbed Xist, that was involved in X inactivation. Further examination of the gene provided yet another surprise in the story of X inactivation. A quick peek back at the Genetics Primer will remind you that genes themselves do nothing. They are first transcribed by a set of cellular proteins into a long single-stranded piece of RNA. That RNA is then translated by a different set of cellular proteins into a functional protein. Except, for the Xist gene, a functional protein isn't the goal. The end product for Xist is the single stranded RNA, which then blankets the entire length of the X chromosome destined for inactivation.

With a gene in hand researchers could begin to work on just what it was about the X chromosome that left it inclined to suffer silencing. It's a problem that has intrigued researchers since the discovery of the phenomenon. Scientific curiosities such as a piece of X chromosome bound to a non-sex chromosome, which then triggered inactivation of the non-sex chromosome, fueled speculation over the X chromosome's character. One theory, proposed in 1983 by Stan Gartler of the University of Washington, Seattle, and Arthur Riggs of the City of Hope Research Institute in Duarte, California, held that the X chromosome is chock-full of some DNA element that promotes inactivation.

Just what that sequence is remains yet another mystery. In 1998 though, Mary Lyon made an extraordinary proposal: what if that DNA element was the so-called junk DNA? Approximately 97 percent of all human DNA doesn't code for protein and has no known function. These sequences include repeat elements such as Alu and LINE-1 elements, but because they are repetitive and appear useless at first, some scientists coined the term "junk DNA." It's likely that there a quite a few gems among the junk, particularly because these junk sequences are maintained through evolution. If they were truly junk, these sequences would degrade over time.

The LINE-1 elements may be a gem for the process of X chromosome inactivation. LINE-1 elements' only known purpose is to replicate themselves and jump from chromosome to chromosome littering the genome with their presence. Lyon noted that, compared to other chromosomes, the X chromosome is saturated with LINE-1 elements. As she saw it, it was possible for Xist RNA to interact with the LINE-1 elements to spread across the chromosome. In 2000 researchers from Case Western Reserve University Medical Center bolstered Lyon's hypothesis by detailing the LINE-1 elements on the X chromosome.

LINE-1 elements represent 26 percent of the entire X chromosome. The greatest concentration of these sequences is in the region directly surrounding Xist. Even more compelling, in areas of the chromosome where genes never escaped silencing, there was a much greater concentration of LINE-1 elements.

While the LINE-1 elements may be playing a critical role in producing X inactivation, it's unlikely they are the only element that aids in shutting down the X chromosome. In fact, researchers from the Dana Farber Cancer Institute in Boston found that the breast cancer gene BRCA1 stabilizes the Xist RNA on the inactive chromosome. When BRCA1 is missing in female cells, it leads to perturbation of the X inactivation and potentially allows genes meant for silencing to escape.

That inactive X is working at levels far greater than anyone expected. In 1999 Huntington Willard, a geneticist now at Duke University but previously at Case Western Reserve University, and Laura Carrel, from Pennsylvania State University College of Medicine, got a hint that more than just the genes with homologues on the Y chromosome escaped inactivation. Out of 224 genes they had surveyed on the X chromosome they found 34 of them were active. And not one of them had a counterpart on the Y chromosome.

Those findings are just a glimpse of how complicated the situation actually is. In 2005 an international team of researchers completed a comprehensive sequence analysis of the X chromosome, finding only 54 genes out of the 1, 098 genes on the chromosome have a functional counterpart on the Y chromosome. In addition, with roughly 1,000 genes, the X chromosome has the lowest gene density of any chromosome. At the same time, the genes for three hundred genetic diseases have been found on the X chromosome. In other words, the X chromosome accounts for only 4 percent of the entire genome, but it houses nearly 10 percent of all of the recognized inherited single gene diseases. The X-linked genetic diseases are striking because they affect males much more profoundly than females.

At the same time, Willard and Carrel expanded their analysis of X inactivation escapees to the entire X chromosome. Only 65 percent of the genes on the X chromosome were fully silenced. Approximately 15 percent escaped inactivation entirely, and another 10 percent of genes varied in their expression. Some of the escaping genes showed significantly less activity than their partners on the "active" X chromosome. Rather than an all-or-nothing situation, the researchers found the "inactive" X chromosome was comprised of fully silenced, partially escaping, and fully escaping genes. In addition, these escapees change with age.

The researchers examined the expression of these genes in forty women, and none of the expression patterns were the same. Not only do women have strikingly different X chromosome expression patterns than men, they have unique gene expression patterns compared with each other. The unequal expression patterns may explain phenotypic differences between men and women as well as those between women themselves.

The genes that did escape weren't spread evenly along the X chromosome, instead they were clustered together. A large group of these genes clustered around the far end of the short arm of the X chromosome, a region that is evolutionarily the "youngest." Presumably, these genes were added on so recently, roughly 50 million years ago, that X inactivation hasn't yet taken hold. Even so, there are escapee genes in the oldest section of the X chromosome as well.

Willard and Carrel also took the opportunity to observe whether LINE 1 elements were correlated with escaping from X inactivation. While the team found a plethora of LINE 1 elements in regions where most genes were silenced, they couldn't see a correlation between LINE 1 elements and the inactivation or escape of specific genes leading them to conclude "genomic and epigenetic determinants are likely to be more complex than repeat content alone."

In other words, the story of X inactivation most likely has a few more twists in its plot. Somehow, it seems only fitting that the genetic process that leads to such complexity in women shouldn't reveal itself all at once.

When a Gene Won't Silence

From the moment the home pregnancy test comes back positive, expectant parents are plagued with competing emotions: excitement about the impending arrival of a son or a daughter and anxiety about the health and well-being of their unborn child. Pregnant women scarf down folic acid-laden prenatal vitamins, give up the smokes, avoid soft cheeses and lunchmeat, cut back on or eliminate entirely the morning cuppa joe all in an effort to give their baby the very best possible start in life. The birth of a healthy little boy or girl nine months later is greeted with both great joy and no small measure of relief.

That joy can turn to despair if a child fails to thrive. Even more devastating for a parent is when a seemingly normal baby suddenly loses all the motor, language, and emotional gains he or she has made during his or her short life. For the parents of one out of every ten thousand to twenty-two thousand girls, that is exactly what happens. A healthy baby girl makes her arrival, she smiles, coos, sits up, crawls and maybe even walks hitting each developmental milestone exactly on time . . . only to start losing those abilities between the ages of six and eighteen months. This baby girl has Rett syndrome.

Rett syndrome was first described in 1966 by an Austrian physician, Andreas Rett, who noticed two girls in his waiting room with the same developmental problems. The medical community had largely ignored the condition until 1983 when a Swedish team, led by Bengt Hagberg, identified the syndrome in thirty-five girls from three different countries.

Girls with Rett syndrome are born after normal pregnancies and routine deliveries. For a time, they develop normally. But between six and eighteen months of age, things begin to go awry, sometimes quite suddenly. At first, the babies start to show signs of autism: they make less eye contact and begin to withdraw emotionally. They may be slow to sit or crawl. Shortly thereafter, girls with Rett syndrome begin a precipitous decline. They start to lose the skills they've spent their entire lives learning. They lose the ability to speak, start replacing purposeful hand movements with repetitive hand-wringing, clapping, and hand-to-mouth motions. Those babies who don't lose their ability to crawl and walk develop an uncoordinated wide-stepping gait as well as an irregular breathing pattern. Half of the girls experience seizures.

While the baby's head size is normal at birth, in most cases her head growth begins to slow between the second and fourth months of life. Over time, most Rett syndrome girls eventually develop a small head, a condition known as microencephaly. Rett syndrome is a common cause of mental retardation among girls.

Curiously, however, the syndrome is almost unknown among boys. That piece of information proved vital to understanding the genetics behind Rett syndrome. When Hagberg and his colleagues reintroduced the world to Rett syndrome, they had found no instances of Rett syndrome among boys. As a result, they postulated that a defect in a gene on the X chromosome caused Rett syndrome. But rather than a recessive mutation like fragile X syndrome or hemophilia, scientists presumed Rett syndrome was a dominant X-linked disease. Whatever the gene associated with Rett syndrome did, its function must be so vital that while females could develop normally for a time, any males so conceived would simply fail to develop and would die in the womb.

With a reasonable place to look, scientists began in the mid-1980s to search for a gene that could explain why Rett syndrome girls thrived early on only to regress into mental retardation. Also in the mid-1980s, a pediatric neurology fellow, Huda Zoghbi who is now a Howard Hughes Medical Institute investigator at Baylor College of Medicine in Houston, Texas, met her first Rett syndrome patient. She was struck by the patient's sudden loss of abilities and her constant hand-wringing. Later the same week, Zoghbi met another patient who'd been diagnosed with cerebral palsy and who also wrung her hands. Zoghbi found the situation a little too coincidental: if there were two patients like this, she suspected there must be more. A quick review of clinical records unearthed cases of five more girls who wrung their hands, failed to coordinate their movements, and suffered from some spasticity.

From that point on, Zoghbi trained her focus on research rather than clinical medicine in order to figure out what was making these girls regress in such a tragic way. While there was good reason to believe that whatever gene was causing Rett syndrome was on the X chromosome, finding that gene proved a difficult task.

Even though Rett syndrome is a genetic disease, it is rarely inherited. Most cases arise from spontaneous mutations; although many sufferers live to adulthood, women with Rett syndrome rarely have children of their own. As a result, Rett syndrome runs in very, very few families, making it more difficult to find the affected gene. Nevertheless, families with half-sisters who suffered from Rett syndrome proved vital to unearthing the gene. The premise that whatever defect was responsible for Rett was indeed lethal in males also gained some credence by studying the Rett families: some of the sons born into these families suffered from severe brain defects and died during infancy.

Because there were so few affected families to compare, researchers couldn't use standard genetic techniques to isolate the genetic cause. Instead, they had to systematically exclude genes on the X chromosome. By 1998 the tip of long arm of the X chromosome was looking like a safe bet as the culprit. Even so, that left approximately two hundred genes to sort through. A group of mice suffering from developmental arrest and fetal death provided the key to finding the gene. In 1999 Zoghbi, who was at Baylor College of Medicine, and Uta Francke, at the Stanford University School of Medicine, announced they'd found the gene responsible for Rett Syndrome: MeCP2.

The function of the gene they'd found was far from what anyone expected: MeCP2 codes for a protein whose sole purpose is to silence other genes.

Just as a reminder, here is a good place to recall that the genes found in DNA do nothing themselves. They need to be "transcribed" into the single stranded RNA, and from there most of the function of the gene only takes place once the RNA has been "translated" into proteins. Genes that are meant to be turned off are "methylated"—a carbon atom and three hydrogen atoms are attached to individual nucleotides.

When it works the way it's supposed to, the MeCP2 protein binds to certain methylated cytosine nucleotides next to guanine nucleotides—in scientific parlance, CpG islands. Regulatory regions of genes are ripe with these CpG islands. In binding to CpG islands, MeCP2 (methyl-CpG-binding protein 2) recruits other proteins and physically prevents the cellular transcription machinery from binding to DNA. As a result, whatever gene MeCP2 binds cannot be expressed.

Zoghbi, Francke, and their colleagues had stumbled upon a conundrum: How could the absence of a gene that turns off a number of different genes throughout the genome have solely neurological effects? The answers are far from complete, but they increasingly point to a critical role for epigenetics—how the regulation of a normal gene can result in genetic effects.

Most interestingly, researchers have begun to look at the individual mutations in MeCP2 to see how they affect the severity of symptoms among Rett's girls. There are variations in the symptoms that constitute the phenotype of Rett syndrome. While most girls suffer from the classic course of Rett syndrome, some will retain their ability to speak and are referred to as having the "preserved speech variant." Others with even milder symptoms are referred to as having "forme fruste" or the "worn down" variant.

Understanding how such variation might develop requires a close examination of the gene product of MeCP2. The MeCP2 protein consists of several critical components, often called domains. The first section of importance is the methyl-CpG-binding domain (MBD). This area quite simply binds to methylated cytosine nucleotides. The next region is the transcription repression domain (TRD), which brings in other proteins to gum up the works and prevent the gene from being transcribed. Within the TRD is a section referred to as the nuclear localization signal (NLS). This section of the protein makes sure the MeCP2 protein stays in the nucleus of the cell—a rather important property if the protein is supposed to act on the chromosomes housed in the nucleus. Finally, the end portion of the gene is called the C terminus.

Where a mutation in MeCP2 occurs, and the type of mutation it is, play a critical role in determining the severity of Rett syndrome. Mutations in the portion of the gene encoding the NLS result in the most severe cases of Rett syndrome because the NLS section of the MeCP2 protein is required to ensure the protein stays in the nucleus of the cell where it functions. When the NLS section of the gene is mutated or missing, the MeCP2 protein can escape the nucleus and fail to aid in the silencing of genes. Mutations in the portion of the gene encoding the MBD and TRD sections of the MeCP2 protein, which shorten the protein, tend to deliver a more serious phenotype than those that alter an individual amino acid. Presumably this happens because truncating the length of the protein impairs the protein's activity to a greater degree than exchanging one amino acid for another. The least deleterious phenotypes are associated with losing part of the C terminal end of the protein.

In general, patients with the preserved speech variant (PSV) tend to harbor some of the milder mutations. But not always. Some PSV patients carry mutations in the critical NLS portion of the gene. In some ways, it's a fluke that these patients escape the more severe forms of Rett syndrome. As described in the Calico Cat Gene, at the earliest stages of development, female cells inactivate one of their two X chromosomes. Ordinarily, it's a random event, and females are mosaics where roughly half of their cells express genes on the X chromosome from mom, and the other half expresses the genes on the X chromosome donated from dad. Sometimes, however, that 50:50 ratio becomes skewed favoring one X chromosome over another. In the case of Rett syndrome when the skewed inactivation favors the chromosome with the normal copy of MeCP2 rather than the mutated copy, that girl may be spared from the more severe forms of Rett syndrome.

Skewed X inactivation also plays a role in the familial cases of Rett syndrome. Women with X inactivation skewed significantly toward the X chromosome housing the normal copy of MeCP2 show few if any Rett symptoms, but they can have severely affected daughters because the daughters inherit the mutated MeCP2 and end up with random X inactivation. Not all familial cases are the result of skewed X inactivation, however. Some Rett syndrome families arise when one of the parents harbors cells with a normal copy of MeCP2 and cells with a mutant copy of MeCP2 in his or her testes or ovaries: the so-called germline cells. That parent is referred to as a germline mosaic, and he or she is capable of producing sperm or egg that harbor either the normal or the mutated copy of the MeCP2 gene, although every other cell of his or her body carries the normal MeCP2 gene. Germline mosaicism doesn't occur often, but when it does a dominant gene may be passed from parent to child even though the parent is unaffected.

Shortly after the gene for Rett syndrome had been identified, the medical community made a remarkable discovery that was contrary to the assumption that allowed the identification of the Rett syndrome gene in the first place: some boys did have Rett syndrome. The discovery meant that the theory that the genetic defect was lethal in males was either wrong or there was something very unusual about those boys.

These Rett syndrome boys did indeed have unusual chromosomes. Several of the boys had two X chromosomes in addition to a Y chromosome. This condition, known as Kleinfelter's syndrome, is relatively common and affects one out of every five hundred males. But with the extra chromosome, and a normal MeC2P to boot, these boys developed classic Rett syndrome as they experienced inactivation of one of their two X chromosomes just like the girls with Rett syndrome.

An extra chromosome didn't explain all of the Rett syndrome cases in boys. Some of these boys were mosaics. In other words, they not only had cells that had a normal MeC2P but also cells that had a mutated gene. At some point during their very early stages of development, one of the cells in the developing embryo developed a mutation in MeC2P gene, and all subsequent cells from that original cell (the progenitor) carried that mutation. As a result, these mosaic boys with Rett syndrome have a similar genetic makeup as girls who have random X inactivation . . . even though they only have one X chromosome. Finally, a curious example of Rett syndrome in a boy came from a male who had two X chromosomes and no Y chromosome at all. The SRY locus from the Y chromosome—the maleness gene, as it were—had been added to the paternal X chromosome, which also had a mutated MeC2P gene.

While it's clear that MeCP2 gene mutations cause Rett syndrome, researchers have found some boys that have mutations in the MeCP2 gene suffer from mental retardation, features of Angelman syndrome, or psychosis. While the phenotypes of these boys differ, most suffer from moderate retardation. Curiously, none of the MeCP2 mutations—typically in the C terminal regions of the protein—have been found in girls with Rett syndrome. Presumably, these mutations are so mild that they don't cause Rett syndrome.

To date, roughly 85 percent of all cases of Rett syndrome can be explained by known mutations in MeCP2. There may be mutations in the noncoding sections of the gene that have yet to be identified, or another gene, which acts in concert with MeCP2 to silence genes, may have defects that result in Rett syndrome.

Even though mutations in MeCP2 have been categorized to phenotype, it still isn't clear how an absence of MeCP2 causes Rett syndrome. There are, however, some tantalizing clues. Rett syndrome was originally thought to be some sort of neurodegenerative disease. It's been shown that the progressive atrophy of the brain (which is the hallmark of neurodegenerative disease) doesn't happen in the brains of people with Rett syndrome: there is no neuronal degeneration in these cases, even though the brains of Rett patients are smaller than normal. Instead, it appears likely that MeCP2 activity is vital to the maturation of the nervous system. In the mouse, MeCP2 is active throughout the brain and other tissues. During times of neuronal maturation, the level of MeCP2 expression markedly increases. MeCP2 activity spreads from the deepest neurons to the most superficial neurons. The timing of this activity mimics the maturation of the nervous system. The rat, too, shows a similar pattern of MeCP2 expression, and there is some evidence that MeCP2 plays a similar role both in rats and humans. In the deepest structures of the human brain, MeCP2 expression is pretty much constant after thirty-five weeks of gestation. But if you only consider neurons in the cortex—the section of the brain housing neurons responsible for speaking, thinking, remembering, and making purposeful movements—you can see a significant increase in MeCP2 expression in the cortex of a ten-year-old human compared to that of a three-month-old human.

The mystery remains how MeCP2 enables nervous system maturation. In 2005, Zoghbi and colleagues revealed the silencing protein appears to play a dual role. Without question, MeCP2 regulates critical gene expression in the nervous system by silencing certain genes at critical times. However, MeCP2 protein also interacts with key proteins involved in processing RNA transcripts. Many genes exist as sections of DNA that provide the code for making proteins interspersed with "noncoding" sections of DNA. When that DNA is copied into RNA, cellular proteins must excise these noncoding pieces before the RNA can be used as a template for making proteins. This operation is called splicing.

Surprisingly, nature makes the most out of the information stored in some genes by using alternative ways to splice together coding portions in order to create different versions of proteins. The nervous system, in particular, makes extensive use of alternative splicing in order to regulate gene expression. MeCP2 appears to participate in this splicing. Zoghbi even goes so far to speculate that MeCP2 might regulate the splicing of the genes it normally silences.

There is still much to be learned about MeCP2. Understanding how neurons develop and function may lead to critical insights into not only Rett syndrome but also schizophrenia, autism, and other neurodevelopmental disorders. The irony of the discovery of the MeCP2 gene is that when this gene that silences other genes fails, the human is left silent.

chapter Five

Just a Little Piece of the Puzzle

Most genes don't individually lead to any particular trait. We don't have a single gene for athletic prowess or behavior, for example. That makes the many genes involved in such traits much harder to identify; scientists can't just compare the genomes of people with certain characteristics like intelligence or athletic ability and then identify a single gene that determines whether you're Stephen Hawking or Steven Segall.

Without a doubt, genetic makeup plays a role in characteristics such as math proficiency and musical ability. But it clearly isn't the predominant force in whether you play Rachmaninoff or chopsticks on the piano. Your drive to practice, exposure to music, and whether your family owns a piano all contribute to your musical success or lack thereof. What follows are stories highlighting genes that play some small role in the complex story of things like heart disease, athletic performance, and behavior. The results of inheriting these genes may not be as dramatic as inheriting the Huntington's disease gene, but they do represent the way most genes exert their influence.

Speaking with a "Forked Tongue"

I can hear my seven-month-old daughter engaging in a one of her endless strings of *ah-da-da-daaas* followed by *mum-mum-mum-maaas*. I'll admit that her easy, incessant experimentation with language has provided hours of amusement for her doting parents and grandparents. Her constant vocalizations have also been the source of some moments of consternation for the unfortunate few who've been subjected to her occasional studies in volume. A few dirty looks aside, her endless mimicry and vocal rehearsal are the means by which she and all other infants learn language from word pronunciation to grammar and syntax.

As much as her family encourages her babbling through "educational" play, my daughter's language ability, like every infant's, is in part an inherited trait. And surprisingly, such ability represents a trait that can be profoundly affected by a single gene.

Whether language and grammar are biological urges, ergo genetic, or the outcropping of social interactions has been a hotly debated topic since linguist Noam Chomsky postulated genetic roots for language in 1959. Chomsky argued that children learn language without express teaching from the adults around them, even though oral communication is very complex to learn. What's more, as a universal human characteristic, many linguists believe language is a likely product of our genes.

Proving Chomsky's assertion has been nearly impossible. While most children easily acquire language, some hit bumps in that road to learning and have trouble with pronunciation, syntax, grammar, recalling words, forming complex sentences, and using their mouth and throats to make meaningful sounds. Linguists, psychologists, and speech therapists refer to this panoply of problems as specific language impairment—a large umbrella term covering language problems that don't have another obvious cause like deafness or mental retardation. It's clear that these types of problems run in families, but teasing out whether the speech difficulties were propagated through social mimicry or genetics hasn't been easy, in large part because the inheritance patterns discovered to date are so complicated. Such complexity is the hallmark of needing many genes in concert with environmental exposures to have a specific effect.

That's where the "KE" family enters the picture. In 1990, researchers from the Institute of Child Health in London described four generations of a large family with a particular speech and language problem. Affected members of the family struggle to control their lips and their tongues—the lower parts of their faces barely move—and speaking even short sentences requires enormous effort. They have problems forming words, comprehending word inflections, and using grammar. They possess obvious deficits in verbal IQ but display nonverbal IQs close to the population average. More important and surprising than the specific symptoms of the disorder is its inheritance pattern: half of the males and half of the females in the family manifest the speech problem.

The language disorder in the KE family follows the classic pattern for an autosomal (not located on the X or Y sex chromosomes) dominant gene. While it was clear that a single gene was the culprit, researchers had no idea where that gene would be found and what it was likely to do.

In 1998, University of Oxford researchers, led by Anthony Monaco, joined Institute of Child Health researcher Faraneh Vargha-Khadem in narrowing down the search for the causative gene to a region on chromosome 7 and referred to it as speech-language disorder 1 (SPCH1). The only problem was there were approximately seventy other genes in that area as well.

The discovery of an unrelated patient (CS) with the same speech disorder helped the teams to pin down the causative gene itself. Two of CS's chromosomes swapped DNA in what scientists refer to as a "balanced reciprocal translocation." A reciprocal translocation is simply the interchange of genetic material between two different chromosomes. For example, a chunk of DNA from chromosome 4 switches places with a chunk of chromosome 8. A balanced translocation means no genetic information goes missing. With no loss of genetic material, persons whose DNA harbor a translocation usually suffer few consequences.

Their offspring, however will either be normal, carry the translocation, or have deletions or duplications of genetic material.

In the case of CS, chromosome 7 was one of the chromosomes involved in the swap, and the swap took place in the gene that causes the KE family's speech problems. The translocation disrupted that "language" gene and caused CS to suffer the same language deficits as the KE family. Monaco and Vargha-Khadem's team used the comparison between the KE family and CS to identify the gene responsible.

What they found is a gene called FOXP2 which belongs to a family of genes known as the forkhead/winged helix (FOX) family. Many FOX genes play crucial roles in the development of embryos. Members of this gene family have been implicated in a number of congenital conditions including congenital glaucoma (increased pressure in the eye) and thyroid agenesis (a condition where the thyroid fails to develop).

All FOX genes encode transcription factors—proteins whose sole purpose is to stick to specific regions of DNA and tell the cellular machinery to activate other genes. A quick glance back at the Genetics Primer will remind you that the DNA that comprises genes doesn't do much by itself. Instead, the DNA must be transcribed by cellular machinery into messenger RNA—a single-stranded molecule far more ephemeral and easily degraded molecule than the double-stranded DNA. After a little processing, the messenger RNA then serves as the template for other parts of the cellular machinery to make proteins that, for the most part, are the action molecules.

The British team found evidence that this particular transcription factor may be important to the developing brain: FOXP2 protein is found in abundance in fetal brain tissue indicating that the gene may play a critical role in the development of normal brain circuitry. In addition, the mouse version of the protein is also expressed in the cerebral cortex of mouse embryos.

Monaco and Vargha-Khadem discovered the KE family all had a single amino acid change in the region of the FOXP2 protein that sticks to DNA, and all members of the family harbored one defective and one normal gene. Because FOX genes are critical during embryonic development, the researchers speculated that the single copy of the normal gene couldn't provide enough of this transcription factor during this critical point in brain development. Scientists refer to this as "haploinsufficiency."

Curiously, like the Huntington's disease gene, FOXP2 gene has a stretch of triplet nucleotide repeats—three building blocks of DNA repeated over and over again—that code for a repeated stretch of the amino acid glutamine in the final protein. The FOXP2 gene harbors a stretch that codes for forty consecutive glutamines. Unlike the Huntington's disease gene, however, the FOXP2 gene's triplet repeats are stable and don't grow in number from generation to generation. One reason may be these nucleotide repeats include two different triplets encoding glutamine—CAG (cytosine, adenine and guanine) and CAA (cytosine, adenine, adenine)—whereas the triplet repeat in the Huntington's disease gene comprises only CAG. This difference may eventually give researchers a handle on understanding what role long stretches of glutamine play in normal proteins.

While the KE family cemented a role for genetics in language, much controversy over that role remains. Some would argue the gene isn't a language gene at all but a gene that controls fine orofacial motor development; after all, the affected KE family members have significant problems using their mouths and tongues. Others would argue that affected patients also have trouble with grammar and semantics, which indicates the gene has at least some effect on the brain circuitry vital for the complexity of language.

The discovery of the gene opens up avenues for more research. Finding the genes that respond to activation by FOXP2 may aid in the identification of other important language genes. While we often consider language to be a uniquely human characteristic, comparing what role FOXP2 plays in other animals will be critical to understanding its role in humans. To date, it appears to play a role in vocalization among songbirds such as finches, canaries, and hummingbirds, as well as mice.

Researchers at Duke University and the Max Planck Institute, led by Erich Jarvis of Duke and Constance Scharff at Max

Planck, analyzed the expression of the bird version of FOXP2 and found that the protein is expressed in the same area of the bird brain as in humans: the basal ganglia. In addition, they found the levels of FOXP2 increase during those times that the bird is learning new songs.

Studies in mice engineered to lack either one or both functional copies of their FOXP2 gene implicated the gene in certain rodent vocalizations as well. Joseph Buxbaum of Mount Sinai School of Medicine led a team that engineered the mice and measured the infant rodents' ability to emit ultrasonic vocalizations (USVs) when they were separated from their mother and littermates. Mice lacking functional FOXP2 protein entirely—called knockout mice—failed to vocalize at all, while the mice that had to rely on the protein produced by a single copy of the FOXP2 emitted many fewer USVs than those with two intact copies of the FOXP2 gene. In addition, the knockout mice lacking any active FOXP2 suffered severe motor-skill impairment and premature death. Mice with a single copy of functional FOXP2 experienced fewer problems but still suffered some noticeable developmental delays.

The Mount Sinai researchers note that other studies of the mice indicate that interfering with FOXP2 affects the migration or the maturation of neurons in the development of the cerebellum, a part of the brain responsible for maintaining posture and balance, and coordinating voluntary muscular movements. Whatever role FOXP2 plays in language and brain development, having an animal model of the disorder will allow scientists to make important inroads.

Undoubtedly, people differ widely in their linguistic prowess. For example, babies learn to speak at very different ages; some people have a gift for grammar; some readily pick up new languages. The full spectrum of language ability inevitably encompasses many genes, most of which have small incremental effects. The FOXP2 gene is one of the low-hanging fruits that now serve as the starting point in the search for understanding how genes have influenced the development and use of human speech and language.

The Cheeseburger Gene

Every one of us who has ever waged a battle of the bulge knows at least one of those chosen few who seem to be able to eat everything that isn't nailed down and doesn't bite back, yet still have trouble *maintaining* their weight. I firmly believe that my best friend since kindergarten is one of them, although she swears her svelte, petite body is the result of daily exercise and healthy eating habits. If I'm to believe her, she splurges on ice cream and french fries only when I'm around.

As far as I'm concerned, this is just another case of genetic inequity. My friend and I could eat the same number of cheeseburgers, exercise for exactly the same amount of time, and I'm pretty certain I'd gain weight while she would continue to bemoan the fact that the Eddie Bauer store near her home recently stopped stocking size two jeans. Her metabolism is designed to burn it, and mine is designed to store it.

My theory isn't without merit. The obesity epidemic that has exploded in the past thirty years is most directly caused by an abundance of food and a paucity of exercise (something obesity researchers refer to as an "obesigenic" environment), but genetic makeup clearly influences how much weight a person gains from such excesses. Studies of twins who were over- or underfed demonstrate that people gain weight when they consume too many calories and lose weight when they ingest too few calories. That result isn't much of a surprise. How much weight the participants in the studies gained or lost varied widely, but pairs of twins gained or lost similar amounts of weight. Because twins share the same genes, researchers concluded that genes help determine how much weight is gained in an environment that promotes weight gain and how much is lost in a weight-loss environment.

Scientists are making headway in understanding how genetics influence the battle of the bulge. The discovery of leptin in 1994 among a strain of grossly obese mice breathed life into efforts trying to link genetics and obesity, at the same time raising hopes that a new class of weight-loss agents would soon become available. In 1999, another weight-loss protein, ghrelin, was discovered. But both of these proteins largely act on the brain by controlling appetite and the sensation of hunger or fullness. Neither gene acts by ramping up the body's metabolic engines. If some people simply burn more calories, some genes must exist that encourage energy to be dissipated as heat rather than stored as fat.

The first candidate for such a gene is a member of a class of proteins called uncoupling proteins and comes from the ultimate low-energy animals—hibernating bears. These uncoupling proteins were identified independently in the mid-1970s by Australian and French researchers. Assuming a hibernating animal would produce increased levels of any proteins that aided efforts to stay warm, the French group, headed by Daniel Ricquier at the National Center for Scientific Research in Paris, compared lab rats kept in the cold to ones kept in warmth. In doing so, the team found uncoupling protein 1 (UCP1) in the animals' brown fat.

University of Dundee biochemist David Nicholls and his team tagged the cellular compartments known as mitochondria as the source of the heat dissipated by brown fat. Mitochondria synthesize ATP—a molecule that stores chemical energy and serves as the body's fuel—from the foods we eat. In order to create ATP, the mitochondria must set up an electrochemical gradient similar to a battery by pumping positively-charged hydrogen atoms across a membrane separating an inner chamber of the mitochondria from an outer chamber. Nicholls's group found the inner mitochondrial membrane of brown fat cells leaked the positively charged hydrogen atoms and prevented the mitochondria from producing ATP. Because the chemical energy had to go somewhere, the fat cells released heat instead of making the ATP that fuels cells. The Dundee researchers identified UCP1 as the source of the leak.

While the finding was interesting for zoologists studying hibernation, it wasn't considered pertinent for obesity research: humans, except for newborns, don't have any stores of brown fat. Even though humans do have a UCP1 gene, the body produces UCP1 protein only during the first couple of weeks of life when the neonate burns brown fat to keep warm.

If mitochondrial uncoupling helps keep humans slim, UCP1 clearly isn't a player. However, another uncoupling protein might be. In 1997, UCP1 discoverer Daniel Ricquier, along with Craig Warden of the University of California, Davis, and Sheila Collins and Richard Surwit of Duke University, identified the gene encoding another uncoupling protein, UCP2, on chromosome 11 and found that it was expressed in brain, fat, and muscle tissues. What's more, several groups of researchers discovered the gene had a cousin sitting right next to it on chromosome 11 that encoded a third uncoupling protein dubbed UCP3.

The newly discovered uncoupling proteins raised hopes of finding roles for uncoupling proteins in human obesity; whatever these proteins did, they didn't simply signal satiety but most likely ramped up the body's engine. The discovery launched a flurry of scientific activity. In 1998, the team that identified UCP2 took the discovery one step further. They studied the gene in two strains of mice: one strain was prone to gaining weight when fed a high fat diet, the other strain remained thin. The weight-gaining mice produced very little UCP2 protein in response to the high fat diet, whereas UCP2 gene expression increased dramatically in the more svelte rodents. In addition, body temperatures increased in the thin mice.

There are only three ways to expend the energy from food: physical activity, our resting metabolic rate (the calories burned simply as a result of being alive), and producing heat. UCP2 appeared to act by helping the body burn calories as heat rather than storing them as fat. The researchers speculate the protein's ability to generate heat may mean it plays a role in fevers and inflammation.

Collins also analyzed UCP3 in the same fat and thin mice. The UCP3 gene, which is expressed only in skeletal muscle, didn't respond to dietary fat the same way that UCP2 did. UCP2 looked so tantalizingly close to that missing link that would help make the connection between body temperature and weight that researchers began to think about developing drugs to increase the amount and function of UCP2 as a means to manage obesity.

As exciting as UCP2's role in weight management looked early on, its role in regulating total body energy expenditure and body weight has hit a bumpy stretch. A strain of mice entirely lacking UCP2 protein maintained body temperature just as well as normal mice and put the kibosh on the idea that UCP2 was a master energy switch. Still, the data indicates UCP2 may aid in setting the resting metabolic rate, which ultimately has some effect on weight maintenance.

Even more important is UCP2's role in type 2 diabetes, which is an insensitivity to insulin—the hormone that signals the body to sop up excess sugar in the bloodstream. Research by Chen-Yu Zhang and Brad Lowell of Harvard University Medical Center in collaboration with Michael Wheeler's group at the University of Toronto indicates that UCP2 dampens insulin secretion in response to blood sugar levels. Using a mouse lacking the UCP2 genes and unable to produce the UCP2 protein (a knockout mouse), the researchers tested how this mouse would respond to varying levels of sugar in its bloodstream. The knockout mouse secreted far more insulin than a normal mouse in response to the same concentration of blood sugar. Pancreatic beta cells release insulin in response to high levels of blood sugar, and they sense those levels via the amount of ATP their mitochondria can produce. Because mitochondria turn glucose into ATP, how much ATP is in the beta cells serves as a good, albeit indirect, measure of the amount of sugar in the bloodstream. In 2002 the Harvard and Toronto groups knew that UCP2 expression reduces the amount of ATP the mitochondria can produce. Based on their knockout mouse findings, they suspected UCP2 expression interferes with insulin secretion.

In 2005, David Jacobs Jr. and colleagues at the University of Minnesota and the University of Texas Health Sciences Center at Houston took the knowledge that UCP2 played a role in insulin regulation and looked for genetic differences, or polymorphisms, that may help identify people at high risk for developing diabetes. The team discovered a common polymorphism in UCP2, which increased the likelihood that a person would develop type 2 diabetes by increasing insulin resistance.

The polymorphism involves a single nucleotide substitution in the UCP2 gene that ultimately causes the amino acid valine to be inserted into the UCP2 protein where an alanine normally resides. The researchers referred to the two alleles of the UCP2 gene as A when the alanine is inserted into the protein and V when the valine is substituted. They found in a study of 3,684 people that a VV genotype—inheriting a V allele from both parents—increased the risk that a person would develop type 2 diabetes. In addition, the association held for men and women, blacks and whites, thin and obese.

Perhaps the most unexpected and exciting role for UCP2 has been as an antioxidant. Turning food into energy by its very nature produces enormous levels of reactive oxygen species (ROS), a form of oxygen that damages cells. Antioxidants render these ROS harmless. Mitochondria are responsible for a large percentage of the ROS generated inside the cell. By causing leaks in the mitochondrial electrochemical gradient, UCP2 and UCP3 decrease ROS generation and provide an antioxidant defense for the cells.

Uncoupling proteins made a splash with the promise of explaining why some people can eat all they want and never get fat. And, while having the right uncoupling proteins certainly helps by making you get the most out of your resting metabolism, the part uncoupling proteins play in diabetes and providing an antioxidant defense are far more important. I can't blame my inefficient uncoupling proteins for my battle with the bulge and wait for a drug to be developed to alter my fate. That's okay. After all, we have a proven method to hike metabolic rate without any weird side effects. It's called diet *and* exercise.

The Bitter Gene, or The Battle over Broccoli

Ask the parents of any toddler what mealtimes are like, and they're apt to recite a litany of complaints from the enormous mess to the refusal to eat anything green or remotely healthy. Psychologists tell us such behavior is vital to the child's developing autonomy, but exasperated parents sincerely wish such psychodramas didn't have to revolve around the dinner table. While it's true toddlers are among the most obstinate creatures on the planet, their refusal to eat foods such as broccoli may well have more to do with their genes and development than their overall temperament.

It's been said there is no accounting for taste, but since the early 1930s science has been able to account for at least one taste: the ability to taste certain bitter compounds that contain a sulfur atom. Arthur Fox, a chemist for the du Pont de Nemours and Company (now DuPont), was working with just such a chemical called phenylthiocarbamide (PTC) in its crystalline form when he spilled a little in the laboratory. The accident was no big deal for Fox, but one of his colleagues complained of a horrible bitter flavor after inhaling some of the PTC dust. It turned out that some people could taste the compound, but others were either completely oblivious to it or required an enormous amount of it to perceive a bitter flavor. What's more, when Fox and other researchers studied families, they found the ability to taste PTC appeared to follow Mendelian inheritance patterns for a recessive trait: most people could taste the substance, some couldn't, and there appeared to be three people who could taste PTC to every one who couldn't. Tasting, it seemed, was dominant to nontasting.

At first blush, the ability to taste PTC seems a curious trait because no vegetable, fruit, tree, or bush produces PTC. It simply isn't found in nature. A person with the ability to taste PTC will also be able to taste other bitter substances that do occur in nature; and many of those substances are toxic, particularly to the thyroid. In other words, the ability to taste PTC may also prevent you from ingesting a naturally occurring poison. It's pretty easy to see how the ability to avoid poisoning could evolve.

What isn't such a simple thing to discern is why "nontasters" exist at all. Presumably, the trait should have simply died out as nontasters ingested toxic plants and berries with no warning that such foodstuffs were decidedly noxious. In 1939, statistician and geneticist Sir Ronald A. Fisher and colleagues examined the bitter-tasting capacity of chimpanzees and found that both tasters and nontasters existed among the apes, and that the tasters outnumbered the nontasters by about three to one. The similarity of the gene-ratio for humans and chimpanzees was striking, and it indicated that the trait existed in what Fisher called a stable equilibrium. For some reason, nature favored the heterozygote—the person who inherited the ability to taste bitter substances from one parent and the inability to taste them from the other.

As Fisher noted in his *Nature* article on the experiment, "it is scarcely conceivable that the gene-ratio should have remained the same over the million or more generations which have elapsed since the separation of [chimpanzees] and [man]. The remarkable inference follows that over this period the heterozygotes for this apparently valueless trait have enjoyed a selective advantage. . . ."

While the inability to taste PTC seems useless, the chemical is very similar in structure to some of the bitter compounds such as isothiocyantes and goitrin found in broccoli, kale, cabbage, and other cruciferous vegetables. Anyone who couldn't bear bitter flavors would likely refuse to eat such healthy veggies. Perhaps, heterozygotes had the ability to not only avoid bitter poisons but also to enjoy eating healthful foods.

Since its discovery, scientists have continued studying the ability to taste PTC trying to link it to certain blood groups and other known genetic characteristics and trying to determine whether the ability to taste PTC actually influenced food choice. For example, President George H. W. Bush famously told his wife, Barbara, that he was president of the United States, he didn't like broccoli, and he had no intention of ever eating it again. While the remark was criticized by some as offering children an excuse to avoid the green veggies on their plates, was it possible that the president's bitter dislike for broccoli was simply the result of his genetic makeup? Quite possibly it was.

In the mid-1990s Linda Bartoshuk, a taste researcher at Yale University School of Medicine who had been studying taste since the 1970s, discovered that not all tasters were alike. Bartoshuk and her group conducted their tasting studies with the thyroid medication 6-*n*-propylthiouracil (PROP), which is very similar to PTC. They found a proportion of the tasters were in fact "supertasters" who found PROP particularly bitter and unpleasant. Bartoshuk and other researchers described 25 percent of all people as supertasters, 50 percent as tasters, and 25 percent as nontasters. Even without knowing which gene was responsible for the ability to taste PTC, the tasting patterns looked decidedly like it was the result of a single gene that had an additive effect.

While looking for the gene or genes responsible for the ability to taste PTC and PROP, a number of different researchers tested whether supertasters avoided other bitter foods. Adam Drewnowski and colleagues from the University of Michigan found that women who were supertasters were more likely than tasters and nontasters to avoid grapefruit juice, broccoli, green tea, and soy products, all of which contain bitter compounds associated with healthy foods.

Researchers were getting a good handle on how supertasting acuity might influence food choices, but the mechanism by which such a trait was inherited remained hidden somewhere on chromosome 7. In 2003, Dennis Drayna and colleagues from the National Institute of Deafness and Other Communication Disorders announced they had found the gene responsible for the ability to taste PTC and PROP. Not surprisingly, the gene was a member of the TAS2R bitter-taste receptor family, which researchers called TAS2R38 or, more simply, PTC.

Drayna's group identified three different changes in the PTC gene called single nucleotide polymorphisms (SNPs), or changes in a single nucleotide of DNA: for example, a guanine where an adenosine would ordinarily be. All three SNPs identified produced amino acid changes in the taste receptor protein TAS2R38. The researchers grouped this triad of changes into combinations called haplotypes. One haplotype was PAV: the amino acid proline at position 49 on the protein; alanine at position 262; and valine at position 296. A second haplotype was

AVI: alanine at position 49; valine at position 262; and isoleucine at position 296. The third haplotype was AAV: alanine at position 49; alanine at position 262; and valine at position 296.

These SNP changes could account for all the variation in the ability to taste PTC. AVI proved to be the nontaster haplotype or allele, and the researchers found that it represented 47 percent of all the alleles in the Caucasian sample and 30 percent of those in the East Asian sample. PAV proved to be the most common taster allele found in 49 percent of the Caucasian sample and 70 percent of the East Asian population. Three percent of the Caucasian population also possessed the AAV allele, which is associated with tasting.

Surprisingly, early expectations that supertasters would carry two tasting alleles and tasters just one tasting allele haven't born out. While supertasters are more likely to harbor two tasting alleles, sometimes a person with just one tasting allele found PTC so bitter that he or she was classified as a supertaster. Drayna believes personality type and perhaps other genes lead to the supertasting phenotype rather than PTC genetics alone.

Because the AVI and PAV alleles dominated, the researchers wanted to know which allele came first and represented the original form of the PTC gene. They sequenced the gene in six primate species including humans, chimpanzees, and crab-eating macaques among others. What they found was that PAV was the only allele found among the nonhuman primates. Whichever allele conferred the nontasting status to chimpanzees is entirely different from the ones conferring nontaster status in humans. In other words, contrary to Fisher's contention that the heterozygote advantage for this gene was established and maintained before humans and chimpanzees diverged, humans and apes evolved the tasting/nontasting trait *independently* sometime after the two species went their separate ways on the evolutionary tree five million years ago, and humans and chimps don't share a nontasting ancestor.

That fact that the nontasting trait developed independently among primates doesn't answer the question of why the trait developed in the first place. While Fisher may have been wrong about the origin of the nontasting trait, he was correct that some sort of selection pressure was keeping the trait around. Otherwise, primates who couldn't taste bitter substances would have died off long ago. Even though the nontaster alleles render a person unable to taste PTC, Drayna points out that it doesn't mean the protein encoded by the nontasting alleles doesn't function at all. The nontasting allele may function by detecting a different set of bitter compounds. If that is the case, Drayna believes it is possible the nontaster allele confers an advantage to PTC heterozygotes—those who harbor both a taster and a nontaster allele—by widening the number of bitter, toxic substances they can detect.

Interesting evolutionary facts aside, understanding the nature of the trait offers an opportunity to understand human dietary behaviors and health. With the gene in hand, it is now possible to ask whether toddlers are just being obstinate or whether they have a genetic excuse for tossing their veggies across the dining room. That's exactly what Julie Menella of the Monell Chemical Senses Center in Philadelphia did.

For the most part, the amino acid that occupies the position 49 in the PTC protein determines whether a person tastes PTC or PROP. Tasters result when either allele codes for the amino acid proline in that position. Nontasters exist when both copies of the PTC gene produce proteins with the amino acid alanine at position 49. The Monell researchers referred to tasters as PP and PA, and nontasters as AA.

Menella and her colleagues tested both the genotypes and the taste preferences of 143 children and their mothers, and included both African Americans and whites in their study. The researchers tested the children's and mothers' reaction to various solutions of the bitter tasting PROP. Not surprisingly, the PP children were best able to identify the bitter flavor even when a miniscule amount of PROP was in the drink they were consuming. Curiously, the PA children outperformed the PA mothers at the same task with 64 percent of the PA children capable of identifying the lowest concentration of PROP compared to 43 percent of the mothers. In other words, genetic makeup couldn't be the sole arbiter of tasting bitterness. Menella maintains the result isn't entirely unexpected because children aren't miniature adults—their sensory systems don't fully develop until after birth. Babies, after all, are known connoisseurs of sweets, a trait that may ensure they get enough carbohydrate energy to grow but at the same time avoid toxic substances. Menella then wanted to know if the PTC gene played any role in this preference for sweets. Indeed it did . . . for the children.

Taster (PP and PA) children had a much bigger sweet tooth than the nontaster (AA) children. The mothers didn't display such a genetic preference, however. Instead, race and ethnicity played a much bigger role in determining the mothers' preferences for sweets with the African American mothers displaying a greater preference for sweets than the white mothers. Sweet preferences followed cultural factors and taste experiences, not genotype.

This is the good news for parents: Taste preferences aren't immutable genetic forces. The genetic component for tasting bitterness is real, but exposure to a variety of foods appears to prime the sensory system to accept foods a person initially rejects as a child. Menella also points out that salt can temper the bitter flavor of many vegetables, and adding a little salt to the offending stalk of broccoli or brussels sprout may make it more palatable to the picky toddler.

In any event, it appears the dinnertime battle may do more than just display the growing independence of a young child—it may set the stage for more adventurous and potentially healthful eating in the adult. In other words, you may lose the battle but ultimately win the war.

The Schwarzenegger Gene: From Mighty Mice to Hulking Human

In the post-BALCO era, many sports fans have come to look at their favorite record-holders with a cynical eye: an athlete with a body a little too beefy and a tad too strong is probably on steroids or some other performance-enhancing drug. Even the erstwhile Mr. Universe, Arnold Schwarzenegger, admitted to using steroids in his heyday—under doctor's supervision, of course. Gone are the days, it seems, when an impressive physique spoke to hours of sweat at the gym and some innate ability to build muscle.

Then again, maybe not.

Just at a time when we've begun attributing nearly all athletic prowess to doping of some sort or another, physicians in Germany have discovered a child who may innately possess all the help he needs to become the next Schwarzenegger. No, he isn't awash in an overabundance of his body's own steroids. Instead, his body simply fails to tell his muscles to stop growing. That biological deficit allows his muscles to become bigger and stronger than most humans'.

This muscle-bound toddler is in some ways the acme of story that starts in 1997 with a curious strain of laboratory mice and a couple of breeds of super beefy cattle, and could end with treatments for muscular dystrophy, AIDS-wasting syndrome, and unfortunately another, perhaps genetically based, means for the doping of athletes.

The story begins with three scientists from Johns Hopkins University School of Medicine on a quest to identify new members of a group of proteins that regulate the growth and development of embryos. The protein they focused on, named "growth and differentiation factor 8 (GDF-8)," is produced at high levels in skeletal muscle.

Se-Jin Lee and his colleagues at Johns Hopkins decided to explore the role GDF-8 played in the body by genetically manipulating mice to lack the ability to produce any of the protein at all. These knockout mice were surprisingly normal, with the exception that their hips and shoulders now carried significant bulk. Lee and his colleagues discovered that these mighty mice put on two to three times the muscle mass of the average mouse.

How could the lack of a gene result in these mouse behemoths? The scientists concluded in 1997 that in normal mice, the protein puts the brakes on the development of skeletal muscle. They named the protein myostatin (in Greek *myo* means muscle, statin derives from the word *statos,* to stop.) The knockout mice had no means to stop their muscles from growing, so they just got bigger.

Perhaps the greatest surprise about these mice was the simple fact that while entirely lacking myostatin, they were in other ways normal and healthy. Assuming that having more muscle and being stronger is a desirable trait, why weren't such mice found in the wild? Lee speculated that the because the mice are a somewhat less timid and a little bit slower, natural myostatin knockout mice may be at a disadvantage in the wild despite their strength.

Nevertheless, the Johns Hopkins researchers' results started a hunt for myostatin mutations in other animals. By comparing the genetic sequences of myostatin in sheep, pig, chicken, turkey, cow, and zebra fish, Lee and his colleague Alexandra McPherron found the myostatin genes were very similar among vertebrate species. Myostatin's high level of similarity between species led the researchers to believe it might play a similar role in other animals.

That's where the efforts of cattle breeding programs in Belgium and Italy came in handy. During much of the last century, Belgian farmers bred cows for a flourishing dairy industry. In the 1950s, Belgium farmers branched out to find ways to breed cattle that provided superior beef. What they came up with was the Belgian Blue. These animals, when fed the same amount of food as normal cattle, developed 20 to 30 percent more muscle than normal cattle. The breed, which represents more than half the cattle in Belgium, is known for its prodigious "double muscling" development. Using entirely different breeding stock, farmers in the piedmont region of Italy had developed their own double-muscled breed, known as Piedmontese, in 1886. Both the Belgian Blue and the Piedmontese cattle are prized for producing meat that is lean and extremely tender.

All that tender meat isn't without cost. Double-muscled cattle often produce offspring that are so large veterinarians must perform caesarean sections on the animals to ensure the survival of both the mother and calf. Some of the animals grow so big they struggle to walk, but otherwise most double-muscled cattle are normal.

Cattle bred to produce hulking musculature is an ideal place to look for natural knockouts like the mighty mice they'd already bred, so Lee and McPherron set out to examine the myostatin genes from Belgian Blues and Piedmontese cattle. Other researchers were of the same mind and beat the Johns Hopkins team to it. Researchers from the University of Liege in Belgium discovered that Belgian Blues lacked myostatin protein entirely because they carried small deletions in both copies of their myostatin genes. Researchers from the United States Department of Agriculture and colleagues from New Zealand discovered the same mutation in the Belgian Blues as well as a single point mutation in the Piedmontese that has much the same effect. The Johns Hopkins researchers ultimately confirmed these results in late 1997.

A little scientific competition is always fun to watch, but what this race actually accomplished—in addition to allowing the North American Piedmontese Association to advertise its cattle as the "Myostatin Breed"—was to establish that myostatin stymies muscle growth in normal animals. As a result, researchers began to question whether myostatin knockouts could be developed for chickens, turkey, pigs, and other food animals to yield more meat. Could blocking the activity of myostatin prove a useful therapy for muscular dystrophy? What about people suffering from AIDS-wasting syndrome where muscles seem to melt away? Would athletes abuse such a drug? And would a naturally occurring human knockout be found?

The answers to most of these questions appear to be a resounding yes. Animal breeders are working hard to develop food animals with less myostatin activity and more muscle.

Researchers from the Washington University School of Medicine in St. Louis and the Charles R. Drew University of Medicine and Science in Los Angeles found that HIV-infected men suffering from AIDS wasting syndrome, or cachexia, had high levels of myostatin expression—their bodies were making large amounts of mRNA, which serves as the active template for making myostatin protein—compared to men who weren't infected and HIV-infected men who weren't suffering from cachexia. In addition, the men with high levels of myostatin mRNA also had low levels of protein synthesis, which is key for developing and maintaining muscle.

Further studies in mice by Johns Hopkins scientists pointed to myostatin's role in cachexia by showing mice that produced too much myostatin lost 33 percent of their total body weight even though they ate normally. As a result the researchers noted developing drugs to inhibit myostatin may prove to be the first effective treatment of wasting syndromes.

Myostatin may also play a role in Duchenne's muscular dystrophy even though the genetic defect causing this form of muscular dystrophy (defects in both copies of the dystrophin gene if you are female and one copy of the gene if you are male) has nothing to do with myostatin. Lee, McPherron, and colleagues found mice with genetic defects in their dystrophin gene were stronger and had more muscle if they also lacked the myostatin gene. This finding is particularly exciting for the muscular dystrophy community because it offers a potential means to treat the disease without relying on the tricky and unpredictable process of gene therapy. The possibility is so enticing that the drug company Wyeth-Ayerst has begun testing a monoclonal antibody against myostatin as a possible therapy.

While new discoveries have abounded since the myostatin gene was identified, without proof that, like the knockout mice, a human lacking active myostatin was normal and healthy, researchers couldn't be certain that efforts to inhibit myostatin wouldn't severely backfire. Humans aren't rodents, and oftentimes strategies that work in mice fail spectacularly in people. Seven years after the discovery of myostatin, German researchers made their announcement about the four-year-old boy who possessed from birth unusual strength and muscle mass. The toddler's body made little or no active myostatin because a single nucleotide switch of a guanine to an adenine caused a fatal flaw in the messenger RNA for the myostatin protein.

The little boy's mother was a professional athlete who carried one gene incapable of coding for a mature myostatin protein. It is possible that as a carrier she produces less myostatin than someone who has two genes encoding mature myostatin protein; as a result she enjoys innate athletic ability. To date, any such assertions are purely speculative. That, unfortunately, isn't likely to deter some from trying to use myostatin genetics to identify more athletically endowed children and steer them toward certain sports.

It also opens up the possibility that today's athletes may try to squelch their own myostatin production with myostatin inhibitors in order to gain an advantage over their competitors. There are, in fact, a few nutritional supplements available that claim to act on myostatin, but there is absolutely no evidence that these products actually deliver.

The discovery of the myostatin gene may indeed help explain in part why some athletes are just a little bit stronger than their peers who train just as hard. Athletic ability has always been a combination of innate ability, will, and training. Just as some athletes in the 1970s and 1980s chose to use steroids to enhance their performance, some will probably try to inhibit myostatin in this new century. Whether that is prudent or fair will be determined by more research and the decision of the World Anti-Doping Agency.

While these issues are important, they are dwarfed by the needs of patients who suffer from muscular dystrophy, AIDS-wasting, and age-related cachexia. For these people, the discovery of the myostatin gene and the potential to develop drugs aimed at squelching its expression offers something that, until recently, was in very short supply: hope.

A Performance Gene

All it takes is one look at the pumped-up Michael Johnson standing next to the reedy Gezahegne Abera to know these two elite athletes excel at different sports: Johnson has had a spectacular career in sprinting, winning three Olympic gold medals and setting a world record, while Abera is the first man to win both Olympic gold and World Championship gold in the marathon. Even though hours in the gym and on the track and the demands of the two sports contribute to the two men's physiques and successes, their genes may play a role as well.

We inherit such physical attributes as hair color, eye color, and body type from our parents, so it's not surprising that our genes would in some way contribute to our physical strength, endurance, and athletic prowess. What is surprising is that a gene encoding an enzyme involved in raising blood pressure grants a strength and endurance advantage. Depending on which version of the "angiotensin converting enzyme," or ACE gene a person harbors, distance or strength events may be that individual's forte.

Like forcing water through smaller and smaller holes to boost the power of a showerhead, ACE raises blood pressure by causing the muscles lining blood vessels to contract and narrow the route through which blood flows. In fact, physicians treating high blood pressure and heart failure routinely give their patients drugs that trump ACE activity in order to open up blood vessels and ease blood flow.

Because of this ability to make smooth muscles in blood vessels contract, Hugh Montgomery of University College, London, wondered whether ACE had any effect on other muscles like the skeletal muscles that move arms and legs, and, if it did, whether ACE activity influenced athletic ability. Montgomery decided to focus on the genetics of ACE.

The ACE gene comes in two varieties: the I (for insertion) allele and the D (for deletion) allele. As a result, an individual may harbor one of three different combinations of these alleles—they may have two copies of the I version, two copies of the D version, or one of each.

In order to study what role, if any, the type of ACE gene had on endurance, Montgomery examined elite runners and tested which versions of the ACE gene they carried. The middle-distance and long-distance runners were more likely to have two I alleles while two D alleles predominated in the sprinters.

The results were far from conclusive, but they did offer a clue that the specific ACE gene configurations one inherits could influence a person's athletic ability. Hoping to discover whether two I alleles of the ACE gene increased endurance, Montgomery decided to look at athletes who routinely tax their endurance—mountaineers who'd climbed beyond 23,000 feet without using oxygen—and compare them to the general population. He found the I allele was particularly common among the mountaineers. In fact, the double I allele was particularly common; but even more interesting, none of the mountaineers carried the double D configuration of genes.

If the I version of ACE really allowed people to exercise more on less oxygen, Montgomery figured he could measure it. Because conditioning and exercise regimens greatly influence physical ability, Montgomery and colleagues looked to fresh military recruits for the answer. The researchers tested the recruits before they started their ten-week basic training course and after they finished. Because the recruits had experienced the same conditioning regimen, the researchers were able to get a clearer look at whether stamina was linked to the ACE gene.

It was. And the I version proved beneficial to endurance activities once again. Recruits harboring two I alleles dramatically increased their ability to curl a thirty-pound barbell repeatedly. In fact, these recruits had eleven times the improvement of their compatriots with two D alleles of the ACE gene.

The association of the I allele with endurance gained support in a 2005 study of mountaineers. Kyriacos Eleftheriou and colleagues at the University College London and the University of Glasgow studied 235 men attempting to ascend Mont Blanc, the highest peak in the French Alps at 15,800 feet above sea level. The researchers analyzed the men's genotypes and matched that against the success each climber had in reaching the summit. Most of the men conquered Mont Blanc. Eight-seven percent of the men with two D alleles and 94.9 percent of the men who carried an I and a D allele succeeded. However, *all* of the individuals harboring two I alleles made it to the top.

But what of the D allele? If the I allele confers an endurance ability, are individuals harboring the D allele destined to be couch potatoes? Actually, no. Jonathan Folland from the University of Brighton and colleagues studied the effects ACE polymorphism on the response of healthy young men to nine weeks of strength training. The men carrying two I alleles gained the least strength throughout the study period, while those carrying one I and one D allele enjoyed the greatest gain in strength.

While the D allele may confer some strength advantage to young British adults, a collaborative study of elderly people in the United States indicates the D allele enhances the benefits garnered from exercise among the aged as well. A team of researchers led by Stephen B. Kritchevsky from Wake Forest University School of Medicine studied 1,024 adults between the ages of seventy and seventy-nine living in Memphis and Pittsburgh. The researchers questioned the participants over three years to determine how much they exercised and whether they were experiencing mobility limitations.

Genotype alone had no impact on whether these elderly participants developed difficulties walking a quarter mile or scaling ten steps. Those seniors who exercised reduced their risk of experiencing mobility limitations by 27.9 percent compared with seniors who didn't exercise. However, when the researchers compared the genotypes of the people who exercised with each other, they discovered dramatic differences. Among the seniors who exercised, those carrying two I alleles developed mobility limitations at a 45 percent higher rate than those who harbored one or two D alleles.

As intriguing as these findings are, they offer no explanation for why a blood pressure gene would influence strength and endurance.

The I version of the ACE gene actually causes the body to produce less of the ACE enzyme. People with two D alleles have much higher levels of ACE activity, and increased ACE activity has been associated with increased skeletal muscle growth in response to overloading the muscles. For those with the I allele having less of the ACE enzyme somehow causes the muscles to use nutrients more efficiently.

Still, there isn't much of an intuitive explanation for why a variation of an enzyme for blood pressure is associated with beefy sprinters and lithe long-distance runners.

In any event, the ACE gene is far from a global indicator of athletic ability. It is at best a very minor player. In fact, the study of elderly people clearly indicates that exercise benefits us all regardless of genotype. There's absolutely no excuse to avoid athletic pursuits simply because of genetics. After all, Montgomery himself isn't going to let his double D configuration of ACE genes interfere with his dream of one day scaling Mount Everest.

An Aging Gene

In May of 2005, Sally Wadyka declared in the *New York Times* that forty was the new thirty. She was talking about the fact that women no longer need to shear their long, luscious locks once they hit the BIG FOUR-OH because women these days are just "younger." One can argue whether people are truly "younger" or simply deluding themselves. Some Americans will go to farcical extremes whether they resort to Botox® injections or brow lifts to stave off the outward signs of aging.

Whether you blame this aversion to aging on baby boomers intent on staying young or on a supersized celebrity culture, America is without question youth obsessed. While we all seem to succumb to the ravages of aging at different rates, even our best efforts cannot forever stall the advance of time. For some, this has been cruelly accelerated by taking a seemingly normal youth through the aging process in fast forward so that his or her hair begins to gray when he or she is just a teenager. This person's twenties are met with rough, wrinkled skin and cataracts. The source of this premature aging condition is the body's inability to keep its genome in neat order. From what scientists are learning about this extremely rare disorder, we now have new insights into the normal processes of aging.

In 1903, while Otto Werner was a medical student working at the ophthalmology clinic at the University of Kiel, he was introduced to four siblings, all in their early to late thirties. Every one of these brothers and sisters displayed premature graying and loss of hair, cataracts, and thickening and hardening of the skin, which Werner described as scleroderma. In 1905, he presented the cases as part of his dissertation. For the next thirty years, Werner's description of the patients was largely ignored until two American physicians, B. S. Oppenheimer and V. H. Kugel, coined the term *Werner's syndrome* for an inherited disease that rapidly ages patients, starting in their early teens.

Werner's syndrome is an extremely rare autosomal recessive genetic disorder. Since 1916, thirteen hundred cases of Werner's syndrome have been reported worldwide, and one thousand of those cases have been identified in Japan. After a normal childhood, the first sign of trouble appears when the child fails to experience a teenage growth spurt. By their early twenties, these patients' hair has grayed and thinned, their skin has wrinkled, and they've developed cataracts. By the time they reach their thirties, they are fighting battles with osteoporosis, hardening of the arteries, diabetes, and tumors of the pituitary gland, adrenal gland, stomach, and pancreas, as well as melanoma and osteosarcoma. All of these degenerative changes culminate in the prominent eyes and beaked noses characteristic of Werner's syndrome. Most patients succumb to heart attack, stroke, or cancer in their forties. For all the premature aging symptoms displayed in Werner's syndrome, patients almost always retain normal healthy immune systems and escape from age-related neurological problems such as dementia.

Even though Werner's syndrome results from a defect in a single gene, its effects manifest in myriad ways. In other words, the disorder, like Marfan syndrome, is pleiotropic.

On a cellular level, George Martin, from the University of Washington in Seattle, discovered in 1975 that Werner's syndrome patients had an inability to keep their genomes tidy, and as a result, their chromosomes frequently swapped and rearranged their DNA. Indiscriminate and inexact chromosome rearrangements increase the chances that a cell will incorporate a mutation. Chromosome instability, however, doesn't explain why Werner's syndrome patients look the way they do.

The first step to understanding the biochemistry of Werner's syndrome came in 1992 when an international team of researchers found a region on the short arm of chromosome 8 that held the gene for Werner's syndrome. In 1996, Martin and his University of Washington colleague Gerard Schellenberg identified the gene (WRN, for Werner's syndrome) as well as four mutations found in Werner's patients.

Curiously, the gene they discovered was an enzyme that unwinds DNA, called a helicase. DNA is comprised of two long strands wound around each other. In order for a cell to replicate and repair its DNA, the double helix must first be unwound. Rather than unzipping the double helix in its entirety, the WRN helicase unwinds small portions of the double helix and moves along the strands, thus creating the room the cellular replication and repair machinery need to operate. As such, the WRN helicase is very similar to two other helicases associated with genomic instability syndromes: Bloom's syndrome and Rothmund-Thomson syndrome. Unlike those other helicases, the WRN helicase also trims away DNA nucleotides that are misplaced in the new DNA strand. This trimming or "exonuclease" activity was first described in 1998 by Judith Campisi and colleagues at the Lawrence Berkeley National Laboratory.

In 2002, Lynne Cox of the University of Oxford established that the two functions of the WRN gene product were used primarily to restart the DNA replication process when it had stalled. In the absence of the WRN helicase, the cell restarted its replication either by recombining the DNA with that from another strand that was replicating, deleting the offending section, or by stalling the cell's efforts to replicate itself—a process called senescence. The problem with these alternative strategies is twofold: recombining or deleting sections of DNA is likely to introduce mutations in critical genes that can unleash cancer, and shutting the cell down limits its lifespan. In order for a cell to develop the defects associated with Werner's Syndrome, the WRN helicase must lose both its helicase and exonuclease activities.

It's not entirely clear how these defects translate into the premature aging associated with Werner's syndrome and whether the disease mimics the natural aging process. After all, while Werner's patients succumb to some of the diseases of age, they are spared dementia and immune-system decline, both of which are common among the elderly. This discrepancy raises the possibility that Werner's syndrome—being sui generis—provides no useful information for the understanding of normal aging processes. Then again, the genomic instability associated with Werner's syndrome may create a "mutator" phenotype that damages DNA, causing cancer and speeding the process of aging. Finally, stalling cellular replication may play a critical role in normal aging by increasing the number of nonreplicating cells in tissues such as skin and the cardiovascular system.

Scientists have known since the early 1980s that the connective tissue cells called fibroblasts from Werner's patients have a limited life span when grown in cell culture. These cells participate in less than twenty rounds of replication called "doublings" before stalling into a state known as replicative senescence, whereas fibroblast cells from normal individuals "doubled" as many as one hundred times in cell culture. Nevertheless, mutations in the WRN helicase don't limit replication in all cell types. For example, even though T cells will develop a "mutator" phenotype in response to a lack of WRN helicase, their life spans aren't affected at all. These results provide intriguing clues about how mutations in the WRN helicase result in tissue-specific aging.

In 1998, scientists from Geron Inc. established that the ends of chromosomes determined the life span of fibroblast cells in culture. These ends carry a series of repeated DNA sequences known as "telomeres," which act a little like the aglet of a shoelace: they prevent the DNA from unraveling. When they get too short, the cell becomes quiescent or senescent. Because telomeres are so vital to maintaining the ability of cells to replicate themselves, cells employ an enzyme called telomerase to keep telomeres at an optimal length. When the telomeres erode, however, a signaling pathway similar to the one that detects

DNA damage is triggered and stops the cell from replicating. That pathway relies on the presence of the tumor suppressor protein known as p53.

When a cell does stop dividing, its gene expression fundamentally changes as compared to that of a replicating cell. In 1993, Vilhelm A. Bohr, of the National Institute on Aging in Baltimore, characterized those changes and found that senescent fibroblasts from Werner's patients didn't display a different gene expression pattern from normal senescent fibroblasts; they just entered the senescent state far sooner. In 2003 Terence Davis, at Cardiff University, confirmed Werner's syndrome fibroblast cells, like normal fibroblast cells, needed the tumor suppressing protein p53 to enter senescence, thereby bolstering the idea that telomere shortening played a role in stymieing cell replication in Werner's syndrome.

Because different cell types display varying levels of telomerase—the enzyme that maintains telomeres, it's not a great leap to envision a mechanism by which Werner's syndrome causes the aging of some tissues while sparing others. For example, skin fibroblasts maintain a much lower level of telomerase activity than the immune system's T cells. This difference in telomerase activity fits nicely with the observation that Werner's syndrome patients suffer from severely accelerated aging of the skin, but they enjoy normal immune system function.

In 2004, however, David Kipling and colleagues at the University of Wales College of Medicine reported that telomere shortening occurs no faster in fibroblast cells from Werner's syndrome patients than it does in normal fibroblasts. Instead, Kipling and colleagues propose that Werner's syndrome results when an as-yet-unknown signaling pathway joins forces with p53 to send a cell into senescence.

Richard G. A. Faragher, of the University of Brighton, proposes that the pathway to senescence is initiated when the replicating DNA becomes stalled, and the mutant WRN helicase can't get it started again. Stalled DNA replication serves as a stress signal that, in combination with telomere erosion, starts the cascade of changes that result in a senescent cell.

As a result, Werner's syndrome bolsters the concept that replicative senescence is at least one cause of normal aging, according to Faragher. Once a certain number of cells have stopped replicating, aging and cell death are the result. What percentage of cells need to stop replicating in order to trigger accelerated aging remains to be seen. But senescence appears to be a crucial pathway in aging.

Unless we're Peter Pan, we all have to grow up and get old. Even though Werner's syndrome strikes very few people, understanding how it ages its victims so mercilessly has given scientists a glimpse of how normal aging occurs.

chapter Six

In the Beginning . . .

Our genes have been handed down to us from the first bacteria that arose from the primordial soup to the moment of our conception. Those millions of years of evolution shaped our genomes, and as such, they bear the marks of ancient infections, survival strategies, and mass migrations. Each gene that was passed on or quieted tells the fascinating story of our genes as they've traversed time.

The Cut-and-Paste Genes

This is the story of a pair of genes that leaped from a germ into a fish and forever changed the course of immunology. As that bit of DNA made its jump, it triggered an evolutionary change: an immune system that tailors its response to the microbial threat it faces. By analogy and in effect, the sophisticated immune system we rely on today may be nothing more than the aftereffects of an infection. That virus or bacterium had sown the seeds of its own doom by endowing us with the raw material to create a new immune system, which can deliver laserlike attacks on invading viruses and bacteria.

Our saga begins roughly 450 million years ago during a geologic springtime: the earth is emerging from one of the coldest moments in its history. The continents are ice-covered and clustered together as a large supercontinent called Gondwana, centered over the South Pole. The climate is warming, melting the supercontinent's thick glacial blanket.

For the first time, coral reefs are blooming pink, ivory, and orange along the coastlines, providing new homes for an explosion of aquatic life. This is the age of fishes, and the oceans are spawning a tremendous variety of them: bony fishes, cartilaginous fishes, and, most significantly, jawed fishes like sharks and lampreys.

Amid this abundance of aquatic life are viruses and bacteria. As is the case for organisms today, some of these sharklike creatures are suffering microbial infections of one sort or another. Whether these fish experienced symptoms of whichever microorganism is to blame couldn't possibly be known, but what we do know is that these creatures' descendants—every animal possessing both a spine and a jaw—owe their existence to that prehistorical ailment because this infection fundamentally challenged, altered, and then improved their immune systems.

It's not that these afflicted fishes lacked all means to battle infection; the jawed fishes and animals as primitive as horseshoe crabs relied on a generalized immune system that recognized some common microbial features and overwhelmed the invading microbes with a host of chemicals and immune cells. The strategy worked then and works today; in fact, it works well enough that the innate immune system is still the first line of defense for organisms scaling the evolutionary tree from horseshoe crabs to humans. This innate immune system defends against invaders with some rather unpleasant and sometimes uncontrollable side effects, such as inflammation and fever. More important, this immunity lacks any means to tailor its response to invading microbes or to remember those previous battles and launch a preemptive strike when familiar foes next attempt to breach an organism's defenses.

Scientists now believe these shark cousins were infected with a microbe that provided a way to create a much more focused, efficient immune system. This particular microorganism housed a pair of genes that like to play the chromosomal equivalent of leap frog. This gene pair comprises a mobile stretch of DNA called a transposon. First discovered in maize, or Indian corn, a transposon is a section of DNA that can snip itself from one place on a chromosome and insert itself into another place on the same chromosome, or on another chromosome altogether. It appears that this particular transposon took an even bigger leap and left the microbial chromosome completely; it ultimately landed in a chromosome of a sperm or egg cell belonging to some of these infected jawed fishes. As a result, that transposon was passed on to all of those creatures' descendents.

This jump meant a DNA free agent had entered the game. And, in an evolutionary blink of an eye, roughly 20 million years, the two genes on this transposon had reinvented themselves and created what is called the adaptive immune system. Those two genes, now RAG 1 and RAG 2, are at the crux of how our bodies launch a sophisticated defense against disease.

This more sophisticated immune system is adaptive because certain immune cells—B cells and T cells—have the ability to fashion proteins to fit specific features of individual invading microbes like viruses and bacteria. Instead of launching an indiscriminate carpet-bombing campaign, the adaptive immune system uses targeted weaponry. The B cells' weapon of choice is the antibody: a specialized protein shaped a little like a two-tined fork that uses the tines to home in on specific threats and neutralize them.

The T cells employ unique proteins coating their surface to identify microbes for destruction. Armed with specialized antibodies and T cells, animals lucky enough to have an adaptive immune system don't often get the same disease twice. If and when they do, the second time around the disease is often much less severe than the first occurrence.

These immune cells face a significant problem, however. The microbial horde harbors billions of unique features, or antigens,

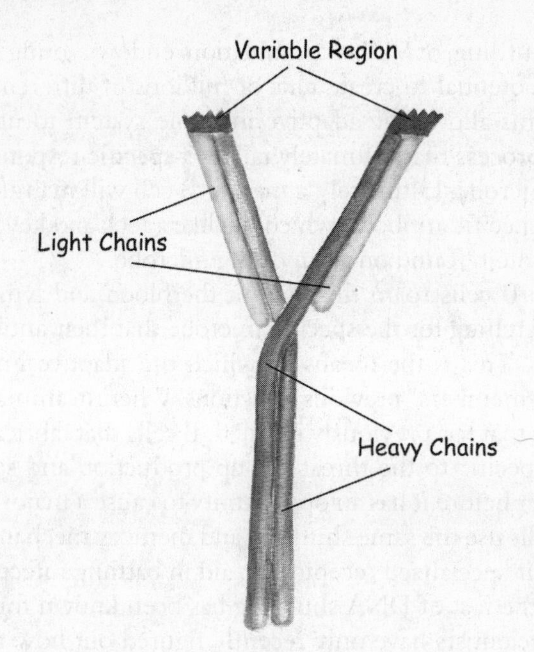

Antibodies are forklike molecules comprised of two light protein chains, two heavy protein chains, and a variable region that can be tailored to recognize specific microbes.

that the adaptive immune system must recognize. If our genomes were to code for all of these possible invaders, the entire length of our DNA would be devoted only to that task. B cells somehow get along with considerably less stored information. Scientists have known since 1976 that B cells accomplish this feat by doing a little DNA shuffling. Instead of encoding a vast array of antibodies, immature B cells make new genes by rearranging existing DNA segments to provide access to an endless library of antibody genes. The B cells randomly swap around three different sections of DNA called V (for variable), D (for diversity), and J (for joining). Once the rearrangement is complete, each mature B cell sports a unique DNA sequence that contains a representative of each of the three sections: V, D, and J.

This shuffling, or VDJ recombination, endows young B cells with the potential to create tens of millions of different antibodies. This allows the adaptive immune system to utilize a selective process that ultimately tailors a specific response to a specific microbe. Ultimately, a mature B cell will manufacture only one specific antibody, which fits like a lock and key with a unique protein found on an invading microbe.

Mature B cells roam throughout the blood and lymphatic system searching for the specific microbe that their antibodies neutralize. This is the means by which the adaptive immune system "remembers" previous infections. When an animal faces a microbe that has previously invaded, B cells that fabricate antibodies specific to the threat rev up production and squelch the invader before it has an opportunity to cause a major infection. T cells use the same shuffling and memory mechanism to create their specialized receptors to aid in battling infections.

While the fact of DNA shuffling has been known for a few decades, scientists have only recently figured out how the recombination takes place. In 1989, David Schatz, a Howard Hughes Medical Institute investigator at Yale University, discovered that the RAG 1 and RAG 2 genes, which sit right next to each other on a chromosome, were absolutely necessary for the DNA shuffling. In 1998, he found out why. Schatz and Martin Gellert from the NIH found that when they put a piece of DNA in between RAG 1 and RAG 2 and mixed it with a piece of circular DNA, RAG 1 and RAG 2 cut the loop of DNA apart and inserted themselves back in, creating a bigger loop of DNA.

In other words, the RAG 1 and RAG 2 genes with the DNA dividing them behaved just like a transposon. This suggested that a mobile stretch of DNA from that ancient microorganism infecting that fishlike creature triggered the development of the adaptive immune system. Because transposons typically leap around the genome, and RAG 1 and RAG 2 appear to stay put, scientists speculated that they must be regulated in a very unusual way.

In fact, a single genetic control near the RAG 2 gene keeps the two hopping genes under control, as Michel Nussenzweig, a Howard Hughes Medical Institute investigator at Rockefeller University, discovered in 2000. Most genes each have their own set of regulatory controls making RAG 1 and RAG 2 rather unusual in the genome. Nussenzweig's finding, albeit esoteric, indicates that the genes were inserted together, not as two separate pieces.

Lost in the focus on proving that the RAG 1 and RAG 2 genes arose from an ancient transposon is the real significance of that event. If RAG 1 and RAG 2 were part of that snippet of DNA jumping from a virus or bacteria to a jawed fish, that one event may have created a quantum leap from a *single* innate immune system to a *dual* system, which comprises the innate immune system as well as the adaptive immune system. Because the newly acquired immune system adapted to the threats it faced and remembered the battles it had already won, animals spent much less energy simply trying to stay healthy. As a result, more complex animals like mammals may have been able to develop. In a sense, without a few sick sharks 450 million years ago, humans may never have evolved at all.

Jomon Genes

Emperor Akihito shocked the Japanese media into silence in December 2001 during a news conference marking his sixty-eighth birthday. He mentioned an eighth-century ancestor, Niigasa Takano, the mother of Emperor Kammu (736–806). Takano wasn't a criminal, crazy, or licentious. She was Korean. When Emperor Akihito mentioned the imperial family's blood tie to Korea, he lifted the veil on some uncomfortable questions about what it means to be Japanese.

The emperor's remarks, the first time anyone in the imperial family had acknowledged Niigasa Takano's Korean heritage, were largely ignored by the Japanese media. In contrast, the South Korean media highlighted and hailed the remarks, noting their Japanese counterparts' reluctance to report on them. The emperor even credited Korean immigration with introducing ancient Japan to another culture and to new technology, and in doing so stirred the debate over how closely related the modern Japanese are to the islands' earliest inhabitants.

As an island nation, Japan takes pride in its unique history and orderly society. But exactly *where* the people who call themselves Japanese came from has been enormously difficult to determine. The controversy over the origins of the modern Japanese surrounds two groups of ancient peoples: The first group, the undisputed original island inhabitants—a unique people called the Jomon, who displayed a curious blend of technological achievements such as pottery making and who led a preagricultural existence; the second group, more technologically advanced immigrants from the Korean Peninsula called the Yayoi. What happened when these two peoples met approximately twenty-five hundred years ago has spawned hotly debated theories over the ethnic makeup of the modern Japanese. Theories aside, genetic evidence increasingly points to a kinship between these two peoples and a Korean ancestry for the modern Japanese.

Japanese culture, for all its technological innovations, is at heart ancient, tradition-bound and, most importantly, insular. This is a country where a *gaijin*—an outside person—seeking to become Japanese and carry a Japanese passport must prove not only that he or she speaks the language but that he or she is "Japanese" enough in his or her habits. Even then, as National Public Radio's Eric Weiner noted, such a person is likely to be referred to as a "gaijin who has a Japanese passport." This ethnocentrism arises in part because the island has been relatively isolated for millennia.

Japan is a country made up of four main islands and strings of smaller islands. From north to south the main islands are Hokkaido, Honshu, Shikoku, and Kyushu. Off the southern tip of Kyushu lie the Ryukyu Islands, a string of smaller islands, with Okinawa the southernmost. The Korean peninsula and the Republic of Russia are situated to the west directly across the Sea of Japan.

The islands of Japan came into existence more than 15 million years ago as a result of motion along the earth's tectonic plates. Japan lies at the intersection of three tectonic plates, the Eurasian plate to the west of the country, and the Pacific and Philippine plates to the east. For millions of years, the Pacific and Philippine plates have been tucking under the Eurasian plate. That's the reason Japan is a seismic hot zone, experiencing more than one thousand small tremors per year in addition to the larger earthquakes that periodically make headlines. Fifteen million years ago, that movement shifted the land masses enough for Japan to break free from the Asian continent.

The islands were totally isolated from human habitation until roughly thirty thousand years ago when the glaciers emerged to form land bridges from Asia to Japan. Twelve thousand years ago, when the waters rose once again to separate the Japanese archipelago, the Jomon were left. Archeologists date the Jomon era from ten thousand years ago to approximately 250 BC. Most Japanese view themselves as direct descendants of these Jomon people.

The Jomon were hunters, fishers, and gatherers living off the fertile streams and temperate Japanese climate. Even though pottery making is usually associated with agricultural societies, the Jomon were, in fact, the first pottery makers. (The Jomon pottery was decorated by rolling or pressing cords into the clay—the Japanese word for cord making is *jomon*.) The oldest Jomon pottery dates to 12,700 years ago, unearthed at an archeological site in the southern island of Kyushu. As the climate warmed, the Jomon appear to have moved north up the island chain.

Around 300 BC, the migration of a group of people from the Korean peninsula interrupted the Jomon way of life. Those people, the Yayoi, sailed to Japan, bringing with them agriculture, weaving, and metal working. Their ability to cultivate rice was the most important technology they shared. Rice cultivation transformed Japanese culture at the time by rapidly taking hold on the islands of Shikoku and Honshu, leaving the colder, less easily cultivated Hokkaido and its unique inhabitants, the Ainu, in relative isolation for the next two thousand years.

There is little dispute that the Jomon were the original inhabitants of Japan. Nor is there much argument that contact with the Yayoi fundamentally altered the Jomon society. In fact, many have pointed to the hunter-gatherer lifestyle of the Ainu (as well as their luxuriant beards!) as an indication that the Ainu are the most direct descendants of the Jomon. If that were true, are the present day Japanese who occupy the central islands of Japan purely Jomon, descendants of the Yayoi, or a mixture of the two? Answering that question may shatter the view of Japan as an ethnically homogenous nation.

One line of thought addresses the interaction between the Jomon and the Yayoi by assuming the modern Japanese are purely Jomon in origins. Supporters of this theory maintain that the Yayoi did indeed come to Japan but arrived in such small numbers that while they provided technological and societal changes, they did not contribute much to the gene pool. One of the most compelling arguments for the this theory isn't genetic, it's linguistic. The Japanese language exists as an outlier with no other language even remotely related to it. As such, some view the language as a derivative of the Jomon language. Had the Yayoi replaced or intermixed with the Jomon, transformation-theory proponents argue that the Japanese would most likely bear some relation to the Korean or Chinese languages.

At the opposite end of the debate, a second theory assumes that most modern Japanese are direct descendants not of the Jomon but of the Korean Yayoi. This "replacement" theory holds that the Yayoi wiped out the Jomon, inhabited the island, and are, in fact, the modern Japanese. This theory explains how the contact with the Yayoi could so quickly change Japanese society.

Most Japanese, however, favor the first theory because it supports the concept of a racially homogenous nation. If the "replacement" theory were true, then the modern Japanese came from Korea. Given the contempt the Japanese have held for Korea, as witnessed by two invasions of Korea by the Japanese in the sixteenth century and Japan's occupation of Korea from 1910 to 1945, it's not entirely surprising the Japanese don't want to see themselves as essentially Korean in origin.

Both of these theories suffer from an absolutist view of Japanese origins: the modern Japanese are either wholly Jomon or wholly Yayoi. In 1991, Kazuo Hanihara, of the University of Tokyo, suggested a third way. Studying the physical characteristics of ancient and modern skulls, Hanihara proposed that the modern Japanese were, to varying degrees, an admixture of both ancient peoples. This "dual structure" model still provides a substantial Korean genetic input into the modern Japanese gene pool. As distasteful as some Japanese may find the notion, genetic evidence increasingly supports this view.

In a sort of ethnic paternity test, Michael F. Hammer, a researcher at the University of Arizona in Tucson, and Satoshi Horai, a scientist with the National Institute of Genetics in Mishima, Japan, looked to the Y chromosome for an answer to the question of Japanese origins.

The human Y chromosome is an evolutionary marvel. Researchers speculate that the Y chromosome, as it exists today, is a shrunken version of an ancestral chromosome that encodes only the necessary genes to make male children. Even though the Y chromosome appears to have been shrinking over time, at some relatively recent point in human evolution the Y chromosome received an addition: an Alu sequence. Approximately 10 percent of all human chromosomal DNA is made up of the short repetitive DNA sequences known as Alu sequences. Not much is known about these DNA elements other than they reproduce themselves and hop around the genome.

The Y chromosome Alu polymorphic element (a.k.a. YAP) made its way onto the human Y chromosome so recently in human history that not all men today have such a sequence. Because a man either harbors a YAP element or he doesn't, it is an attractive way to analyze origins.

When Hammer and Horai first started looking at YAP for some answers, they found that the only Asian men who possessed the YAP marker were Japanese. None of the Taiwanese, Korean, or Chinese men displayed the YAP element. However, not all Japanese carried a YAP marker. The Ainu of the northern island of Hoikkado and the Okinawans and other Ryukyuans inhabiting the southernmost parts of Japan were the most likely to have the YAP element. Japanese men who lived on the central islands were the least likely to have the YAP element.

Hammer and Horai attributed the YAP element's presence to the Jomon people. When the Yayoi came to the central islands of Japan, they brought with them a Y chromosome lacking YAP. The men of central Japan bear the genetic evidence of the Yayoi arrival. Additionally, Hammer and Horai studied a DNA marker on the Y chromosome common among the Koreans, O-SRY. This marker also appears most commonly among the men of central Japan and least often among the Ainu and Okinawans.

Surprisingly, the YAP analysis uncovered a YAP element among Tibetans as well. Because humans migrated out of Africa to populate the rest of the globe, that particular YAP element may have originated in Tibet. As the people who became the Jomon migrated across Asia, this genetic legacy of their travels appears to have died out.

In solving the mystery of the peopling of Japan, mom's mitochondrial input bolsters the argument for the dual structure hypothesis, while at the same time complicating matters a bit. The energy generating mitochondria are unique components of cells that contain their own chromosome and are inherited only from our mothers. As a result, mitochondrial analysis can establish maternal heritage. In 1996, Horai and colleagues at the National Institute of Genetics in Shizuoka, Japan, analyzed several genetic markers from mitochondria. Like the YAP analysis, Horai's mitochondrial DNA analysis indicated that more Japanese living on the central islands carried mitochondrial markers similar to the Koreans and Chinese than either the Ainu or the Ryukyuans. In fact, Horai's work indicated nearly 65 percent of the mainland Japanese gene pool is the result of interactions with the Asian continent after the Yayoi period.

The story became even more complicated with the results of a detailed analysis of the mitochondrial DNA conducted in 2004. Masahashi Tanaka, from the Gifu International Institute of Biotechnology, in Gifu, Japan, and an international team of researchers confirmed Horai's work by finding a significant Korean influence among the mainland Japanese. The Ryukyuans and the Ainu displayed different maternal origins. What's more, the Ryukyuans and the Ainu differ significantly from each other: they have separate maternal origins. Ainu mitochondria displayed some similarity with Siberian peoples, and the Ryukyuans' mitochondria bore the evidence of migrations from southeast Asia.

As for the genetic legacy of the Jomon, Tanaka and colleagues found the Ainu were most directly descended from these earliest people of Japan, while the Ryukyuans were the next closely related, and the mainland Japanese the least closely related to the Jomon.

The genetic evidence now provides a complex picture that details not only the Yayoi genetic contributions but many other genetic influences on the modern Japanese gene pool. While Japanese culture may be having difficulty accepting the genetic heterogeneity of its people, Japanese science continues to map the migrations of ancient peoples to the Land of the Rising Sun.

Survivors' Benefit

Europe and the Middle East had emerged from their Dark Age to embrace culture and learning. Athens was establishing its democracy, and Rome was giving birth to a republic. Homer had written the *Iliad* and the *Odyssey*. In China, the age of Confucius was beginning, and the Silk Road trading route started to link the great cities of Asia to Rome.

At some point during this time of nation building, cultural growth, and exchange, the normal processes of copying DNA went awry for someone in Europe or the Middle East. The source of the mutation may never be known, but it is certain that this mutation manifested itself in the sperm or egg cells of that individual because the genetic defect found in a critical receptor of the immune system spread among this person's descendants.

Too often, the story of a genetic mutation ends in devastation because it's easier to recognize when a DNA alteration wreaks havoc. But this time Providence smiled, and that mutation proved most advantageous: the 10 percent of all Europeans who carry the mutation enjoy resistance to infection with HIV, the virus that causes AIDS.

Still, the prevalence of the mutation remains a conundrum. How is it possible that mutation arose millennia in advance of the debut of the viral scourge it protects against? Why do so many Europeans enjoy some level of protection against infection with HIV as a result?

For the mutation to take hold in such high numbers, it must have conferred some sort of advantage to its carriers. Perhaps the mutation increased fertility, or maybe it protected against coronary heart disease. The most popular explanation, however, is that the mutation thwarted the efforts of some other infectious agent that had its heyday in Europe during the Middle Ages. When it comes to medieval scourges, plague and smallpox top the list. Discerning which of these afflictions favored the mutation has scientists employing mathematical models and questioning historical assumptions about these diseases.

The favorable mutation in question alters a critical protein that HIV needs to enter a human cell. Much like a rock climber scaling a cliff, HIV needs both a toe- and a fingerhold in order to infect the immune system cells it targets. In the mid 1980s, researchers discovered the fingerhold—a protein called CD4 on the surface of the immune cells, or T cells, the virus used to gain entry. In 1996, a number of researchers discovered the toehold—yet another cell-surface protein. T cells roam the body by responding to the presence of small signaling proteins known as chemokines. A chemokine receptor is the means by which T cells sense chemokines and go on the move. HIV gained entry into T cells by grabbing onto the chemokine receptor called CCR5.

That same year, a team from the Aaron Diamond AIDS Research Center at Rockefeller University found people with a 32-base pair deletion in both copies of their CCR5 genes (homozygotes) were resistant to infection with HIV. The mutation, referred to as CCR5-Δ32, removes a chunk of DNA in the middle of the gene and causes the body to produce such a severely truncated protein that no CCR5 can be found on the surface of a homozygote's T cells. Additionally, people with one normal copy of CCR5 and one mutated copy enjoy a 70 percent reduced risk of infection compared to people with two normal copies of the gene.

Scientists began speculating that some other deadly infectious agent bore responsibility for the selective advantage the mutation conferred when, in 1998, an international team led by Marc Parmentier of the Institute of Interdisciplinary Research, Université Libre de Bruxelles, Belgium, documented the frequency of the mutation in the European population. They discovered that northern Europeans were far more likely to carry the CCR5-Δ32 mutation than their southern European counterparts. The mutated gene cropped up in 4 percent of Sardinians, 11 percent of French, and 16 percent of Finns and Russians.

When it comes to medieval scourges, the Black Death, or bubonic plague, it is usually the epidemic that comes to mind—and for good reason. The Black Death wiped out roughly a third of the population in Europe between 1347 and 1350. The pestilence traveled north from Italy, then throughout Europe, killing an estimated 25 million people in its wake. It was 150 years before Europe regained the population it had before the plague. The heavy toll exacted by the Black Death makes for a powerful selective force. For four hundred years, Europe suffered from intermittent plagues, the Great Plague of 1665–66 centered in London killed between 15 and 20 percent of the British population.

Researchers began favoring the theory that Black Death was the selective force driving the increase in CCR5-Δ32 mutation frequency because the estimated allele frequency began to increase in the fourteenth century. That timing matches the Black Death almost exactly.

At the same time, Europe was battling a less dramatic but still deadly scourge: smallpox. While no single epidemic of smallpox killed as many people as the Black Death, repeated smaller outbreaks occurred over the past two thousand years, ultimately killing more people than the plague. Smallpox also has a diabolical affinity for killing children. If the CCR5-Δ32 mutation protected against smallpox infection, children resistant to the disease would more likely survive an epidemic, grow up, and pass the CCR5-Δ32 mutation on to their children.

The smallpox theory received a boost in 1999. Grant McFadden and colleagues at the University of Western Ontario discovered that a pox virus (called myxoma virus), which infects rodents and rabbits, used the CCR5 protein in the same way as HIV to infect immune cells. As a pox virus, myxoma virus is a close relative to *Variola major,* the virus that causes smallpox, thereby bolstering the case for smallpox as the source of selective pressure for the CCR5-Δ32 allele.

The smallpox camp gained further support in 2003 from a mathematical model developed by University of California, Berkeley, population geneticists Alison Galvani and Montgomery Slatkin. The Berkeley team that found the outbreaks of bubonic plague were too rare to have increased the prevalence of CCR5-Δ32 to the levels we see today. Smallpox, on the other hand, provided a steady pressure forcing up the prevalence of the mutation over time. When Joan Mecsas of Stanford University Medical School and colleagues discovered that the bacteria that causes bubonic plague, *Yersinia pestis,* infects mice with the CCR5-Δ32 mutation just as easily as it does normal mice, the plague theory appeared to have suffered a fatal blow.

What if the entire premise that plague is caused by a bacteria is wrong? What if the true cause of Black Death was not a bacterium but a virus that causes bleeding similar to Ebola virus? That's just what Christopher Duncan and colleagues from the University of Liverpool proposed in 2005.

Puzzled by the higher frequency of the mutation among people in northern Europe, Duncan and his colleague Susan Scott maintain that smallpox doesn't account for the prevalence trend, in part because they say a deadly form of smallpox didn't develop in Europe until the 1600s, far too recently to account for the high prevalence of the CCR5-Δ32 allele today. At the same time, the team agrees that a bubonic plague caused by *Yersinia pestis* also couldn't account for the selective pressure that caused the CCR5-Δ32 allele to reach such a high frequency as it had in Europe, mainly because bubonic plague wasn't directly infectious between humans but was spread by rats and their fleas. Focusing on the fact that forty-day quarantines proved particularly successful in stopping the spread of plague, Duncan and Scott propose the Black Death was, instead, a viral hemorrhagic fever that could be transmitted directly from person to person. The black swellings attributed to the characteristic buboes of bubonic plague, while dramatic, weren't the most common harbingers of the plague. Duncan and Scott maintain "God's Tokens," hemorrhagic spots on the chest, throat, arms, and legs, affected the overwhelming majority of plague patients. These symptoms are characteristic of hemorrhagic viral illnesses like Marburg virus, Dengue fever, and Ebola virus.

Viral hemorrhagic fevers have been recorded in the Mediterranean and North Africa for more than three thousand years. The Liverpool team argues that after the single mutation event more than twenty-five hundred years ago, the frequency of the CCR5-Δ32 mutation was forced up to a level of one in twenty thousand by the fourteenth century when the Black Death struck and provided a heavy selection for the mutation. Periodic outbreaks, including the Great Plague of London in 1665, forced up the frequency of the mutation by killing people with normal copies of the CCR5 gene that were susceptible to plague.

Duncan and Scott believe their hemorrhagic fever theory explains the increased prevalence of the CCR5-Δ32 allele in Scandinavia and Russia because outbreaks of hemorrhagic fevers continued to ravage those areas until the 1800s. In addition, because smallpox probably uses the CCR5 protein to infect immune cells, smallpox epidemics might have provided additional selective pressure to keep the CCR5-Δ32 allele at high frequencies after the hemorrhagic fever outbreaks faded into history.

The hemorrhagic fever hypothesis requires a fundamental rethinking of the causes of Black Death. Unless we can find plague victims encased in permafrost and then reconstruct the virus from the preserved body, Scott acknowledges there is no way to establish that the Black Death was the result of a viral hemorrhagic fever. While the hemorrhagic fever theory is intriguing, the difficulty in proving it leaves smallpox infections as the favored hypothesis for explaining the selective force that increased the frequency of the CCR5-Δ32 mutation. Whatever the cause, one thing is certain, the mutation has proven a blessing since it first appeared in antiquity.

The Pregnancy Genes

Lurking deep within our genomes are the legacies of ancient viral high jinks. Scattered about our forty-six chromosomes are more than one thousand regions—up to 8 percent of the DNA in our genomes—where viruses, which once infected some evolutionary ancestor, put down roots in that organism's germline cells and have been inherited ever since. No longer infective, these former viral terrors have been reduced largely to pale imitations of themselves. Most of them are nothing more than dilapidated DNA outposts.

That's not to say all of these embedded viruses have been banished to antiquity. Some appear to have the potential to be active in the cell and yet seemingly cause no mischief. Evidence is mounting that these incorporated viruses may provide such useful services as protecting us from infection with other viruses. The most stunning possible role for three of these vestiges of ancient infections is in the development of the first organ that every mammalian embryo makes: the placenta.

The placenta is the interface between mom and baby. This first organ, comprised entirely of cells from the fetus, taps into mom's blood supply to provide nourishment to the growing baby. While serving as the food and oxygen source for the developing fetus, the placenta keeps mom's immune system at bay while providing critical hormones both early and late in the pregnancy. Developing a placenta was the critical evolutionary step needed to allow most marsupials and mammals to grow their offspring internally rather than hatching them from eggs.

The first hint that an embedded virus may be playing some role in the development of the placenta came in the early 1970s. Electron microscope examinations of the placental tissues from a number of different animals, including baboons and humans, unearthed what appeared to be viruses forming, budding, and escaping from the placental cells. Unless all of those animals were suffering from a viral infection, the particles came from the placenta itself. Those particles were in fact embedded viruses—endogenous retroviruses to be exact—and the placentas were simply teeming with them. In fact, endogenous retroviruses represent a mere tenth of a percent of all protein produced by the placenta.

Retroviruses comprise an entire class of viruses that incorporate into the host genome as a means of replicating. RNA, in a genetic about-face, makes up the genome of these viruses. The term *retrovirus* applies to the fact that these genes reverse the "central dogma" of genetics that holds genetic information is stored in DNA, transmitted by RNA, and realized as proteins. Retroviruses double back on that path by turning their RNA genomes into DNA and incorporating into the host genome in order to reproduce. For example, the human immunodeficiency virus (HIV) is a retrovirus. When a retrovirus incorporates into germline cells, it can become a permanent fixture inherited from one generation to the next.

So far, researchers have identified three endogenous retroviruses as actors in the human placenta: ERV-3, HERV-W, and HERV-FRD. The question is, what are they doing there? Nobody knows for sure, but Eric Larsson, a pathologist at the University of Uppsala in Sweden, proposed in 1988 that endogenous retroviruses play two roles: causing a section of the placenta to fuse together, and suppressing mom's immune system during a critical period of development.

In order to get a handle on how endogenous retroviruses could possibly play a role in human reproduction, we need to take a closer look at the placenta's role in evolution and its structure in humans.

Roughly 130 million years ago, during the time the Alps were being formed, the animal kingdom was making a giant leap. Animals—mammals and marsupials to be exact—were evolving to give birth to live offspring instead of producing offspring that needed to hatch from eggs. Although its origins remain a mystery, the evolution of the placenta was vital to this effort. The placenta is the route through which a fetus taps into the nutritional resources of the mother for the duration of its development in the womb.

Gestating offspring inside the body offers several advantages to developing inside an egg. First, the fetus has a means to eliminate waste rather than stewing in the by-products of its own growth and development. In addition, a fetus developing inside its mother benefits from a ready supply of oxygen and nutrients. These two features may very well have bought mammals enough developmental time and resources to evolve the big brains for which the entire class is known.

While creating an organ that allows for the exchange of nutrients and wastes seems a logical enough way of achieving live birth, it's not without its potential pitfalls. First of all, the fetus is by definition half foreign. Half of its genes come from the father and, as a result, as many as half of the proteins the fetus makes may be different enough from maternal proteins to trigger an immune reaction. If the fetus is going to develop inside the mother, it has to find a way to escape any efforts by mom's immune system to eliminate the "foreign invader." Second, its parasitic existence means the fetus not only gets the nourishing goodies from mom but also has the potential to contract any viruses or bacteria infecting mom. As a result, the placenta must serve as both gateway and shield.

By day three after fertilization, a zygote begins to divide, forming a spherical ball of cells. Around day five, that ball of cells becomes a hollow blastocyst, which has a peripheral layer of trophoblast cells surrounding the inner cell mass. The trophoblast is destined to become the placenta; the inner cell mass will eventually become the fetus.

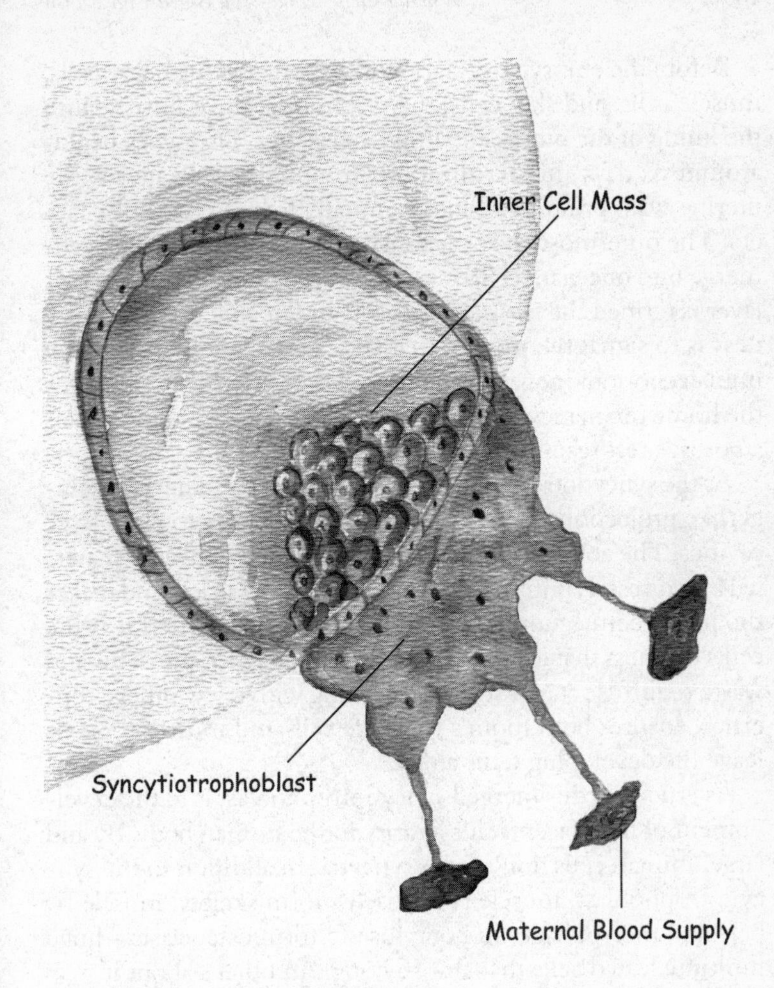

The blastocyst, a fertilized egg that has divided a number of times and turned into a hollow ball, consists of a peripheral layer of cells called the trophoblast surrounding the cluster of cells forming the inner cell mass. Destined to become the placenta, the trophoblast burrows into the lining of the mother's womb and becomes two layers. The cells in the outermost layer merge to form a giant cell with many nuclei called the syncytio-trophoblast, which then taps into the maternal blood supply.

Before the embryo can start working on creating nerve cells, muscle cells, and skin cells, the trophoblast must burrow into the lining of the uterus and build a placenta. That work begins around six days after fertilization. Trophoblast cells invade the uterine wall, proliferate, and eventually form two cellular layers. The outermost layer then takes an unusual turn: The cells merge into one giant cell, or syncitium, with many nuclei. This layer is termed the *syncytiotrophoblast*, and its first order of business is to signal the mother that she is pregnant by producing human chorionic gonadatropin (hCG)—the hormone that turns the home pregnancy test pink or blue or whatever color signals a positive test result for the test in use.

As the syncytiotrophoblast continues to grow, it develops fingerlike projections and erodes the endometrium to form large cavities. The gossamer-thin syncytiotrophoblast invades maternal blood vessels until the blood fills the empty spaces and makes the initial connection between offspring and mother. It is the cell's thinness that leaves it permeable enough for nutrients and waste to diffuse across. Its fusion, along with a few other properties, ensures both mom's immune cells and most pathogens leave the developing fetus alone.

As crucial as this merged syncytiotrophoblast is to the development of the placenta, it's a rarity in the human body. By and large, human cells don't form syncytia. In addition to the syncytiotrophoblast, muscle cells fuse to form skeletal muscle fibers and macrophages in bone fuse to form osteoclasts—huge multinucleated cells that absorb bone. And that's about it.

Retroviruses, on the other hand, encourage the formation of syncytia. And this is where the endogenous retroviruses (ERVs) come in. In order to replicate themselves, retroviruses as a class must insert themselves into the DNA of the cell they are infecting and hijack the cellular machinery to make copious copies for further infection. It's the ultimate free ride. The big payoff comes if the virus can infect an egg or a sperm cell; assuming the host survives the infection long enough to reproduce, the virus is now part of the host's genome and can be transmitted through the generations. Presumably, an evolutionary ancestor to mammals employed some of these embedded viruses to evolve the placenta.

The idea isn't so far-fetched. After all, HIV-infected cells produce envelope protein so that they can fuse with uninfected cells and spread the infection. If this is true in the case of the placenta, it doesn't take much of a leap to surmise the ERVs help trophoblast cells fuse to become the syncytiotrophoblast. In addition, retroviral envelope proteins harbor a section that quiets immune cell activity and may therefore provide protection for the fetus from the mother's immune system.

ERV-3 was the first of its class to be implicated in placental development. Discovered in 1984, it resides on chromosome 7. In 1996, Eric Larsson and colleagues discovered that one third of all pregnant women produced elevated levels of antibodies against the ERV-3 envelope protein compared with only 6 percent of normal nonpregnant women. These elevated levels increased during the length of the pregnancy and plummeted to normal levels after birth, indicating the presence of large amounts of ERV-3 protein only during pregnancy. In 1999, Neal S. Rote, now at Case Western Reserve University School of Medicine, tested whether ERV-3 could induce the merging of cells into a syncytium The team added the ERV-3 gene to a laboratory culture of trophoblast cells. The cells altered their shape and started producing hCG—the pregnancy hormone—in the same way the syncytiotrophoblast does.

While intriguing, whatever role ERV-3 plays in the development of the placenta, it most likely isn't novel or critical. Shortly after Rote and his colleagues announced their findings about ERV-3, Nathalie de Parseval and Thierry Heidmann from the Institute Gustave Roussy in Villejuif, France, documented a rare genetic polymorphism for ERV-3: the gene produces a severely truncated form of the protein consisting of only twenty-five amino acids. In their small study of healthy Caucasians, 1 percent (three individuals) had two of these shortened alleles and failed to produce any normal length ERV-3 protein. If ERV-3 were essential for placental development, these people should have never been born: at the earliest stage of development they would have failed to make a functioning placenta.

While the French group's findings could be viewed as the death knell for ERV-3's role in the evolution of the placenta, Rote suggests several reasons why ERV-3 shouldn't be so quickly discounted. First, the truncated portion of the protein may have some vital activity. Second, even though the individuals identified by the French team clearly survived pregnancy, there is no documentation of whether there were any abnormalities with their placentas, with their mother's health during the pregnancy, or with their birth weights. Another possible explanation for the French results could be that additional ERVs or ERV-like elements take up the slack for the missing ERV-3.

The most likely explanation in Rote's opinion is that ERV-3, which is produced in a number of tissues in addition to the placenta, may aid in trophoblast differentiation as well as playing a generalized role in hormone production rather than any role in fusion or immune suppression. He suggests the twenty-five amino acids produced by the subjects in the French study may be just enough to get the job done.

The scientific community didn't have to wait long for another placental ERV to come to the fore. A group of scientists at Genetics Institute, Inc., in Cambridge, Massachusetts, isolated a retrovirus-like protein sitting on the surface of human testes. When the team comprised of John McCoy, Sha Mi, and James Keith searched for the protein in other human tissues, they found it was even more abundant in the human placenta.

Knowing that other retroviral envelope proteins will fuse cells, the researchers tested how well this newly discovered protein completed the task. Adding the gene to nonplacental cells caused the cells fused into a syncytium. The cells would even fuse with liposomes, those spheres of fatty acids that mimic the cell membrane. In addition, when trophoblast cell culture cells were treated with a chemical that caused them to fuse, the expression of this envelope protein skyrocketed. The researchers dubbed their protein *syncytin*.

Around the same time, a French team headed by Françoise Mallet of the Ecole Normale Supérieure in Lyon, France, identified an endogenous retroviral-like envelope protein, which was active only in the syncytiotrophoblast and could fuse cells in the test tube. They named the protein HERV-W—it was the same protein as syncytin. They also found that HERV-W spurred the production of hCG.

Even more interesting, the two groups asked what would happen if they blocked the action of HERV-W. If the researchers were correct and HERV-W played a critical role in trophoblast fusion, then blocking HERV-W activity should stymie cell fusion. The Massachusetts team tested antibodies against HERV-W and found that few trophoblast cells in culture fused in the presence of the antibodies. The French team employed special RNA molecules called antisense RNA. These antisense molecules bind to the messenger RNA transcript of the HERV-W gene and prevent it from being translated into protein. The antisense RNAs decreased both cell fusion and the production of the pregnancy hormone hCG.

Whereas altered ERV-3 production doesn't appear to cause significant problems during pregnancy, altered HERV-W production may be associated with preeclampsia, or uncontrolled high blood pressure during pregnancy. The Genetics Institute team discovered that patients with preeclampsia had a decreased level of HERV-W expression in the syncytiotrophoblast. Bolstering the Massachusetts team's finding in 2002, Ina Knerr and colleagues at the University of Erlangen–Nuremberg in Germany found reduced HERV-W expression in the placentas of women suffering from preeclampsia and HELLP (hemolysis, elevated liver enzymes, low platelets) syndrome. Still, Neal Rote notes that it is unclear whether the HERV-W expression is the cause or the consequence of these pregnancy complications.

As a role for HERV-W in the development of the placenta appears to be firming up, there is a new kid on the block. In 2003, Nathalie de Parseval and Thierry Heidmann made an attempt to identify all retroviral-like envelope regions in the human genome. In the process they identified a new endogenous retroviral envelope protein. Like HERV-W, HERV-FRD can initiate cell fusion in nonplacental cell lines. In addition, HERV-FRD was able to accomplish the task in cells lines that HERV-W couldn't fuse and vice versa. Such a result indicates that the two HERVs are using different cellular receptors to initiate cellular fusion. Indeed, the French team may have uncovered some of the evolutionary redundancy that both Larsson and Rote anticipated. After all, certain things in life are too important to lack a backup system.

The question comes back to why. Why is the placenta brimming with the evidence of these former infections? Even though scientists have been studying endogenous retroviruses for decades, they still haven't described a physiologic role for them. The story of ERV-3, HERV-W and HERV-FRD opens the door to the first real understanding of their purpose. The ground is becoming firmer for those who maintain these retroviral relics are critical elements in the evolution of that transitory organ that sits at the heart of mammalian success: the placenta.

The "Got Milk?" Gene

For all the celebrities the dairy industry has paid to sport milk mustaches, and all the money they've spent touting the slimming qualities of the calcium found in dairy products, the industry can't spend enough money to get around one simple fact: most adults in the world are lactose intolerant. They simply can't digest the primary sugar in milk.

Lactose intolerance sounds like a disorder—and it does produce significant albeit short-term suffering for the afflicted who make the mistake of quaffing a cold glass of milk—but it is the natural state for most adult mammals, humans included. The unusual, dare I say mutant, persons are those who can still enjoy milk with their cookies well into adulthood.

Scientists have known the full structure and sequence of the lactase gene since 1991. Fourteen years later, they still don't know what causes lactase persistence; there aren't obvious differences in the gene and its flanking sequences between those who can drink milk and those who can't. By analyzing the smallest genetic differences, however, scientists are beginning to get some clues.

All newborn mammals begin to produce high levels of the enzyme lactase within the first few days of life. They must do so in order to thrive because all mammalian milk contains lactose—a compound sugar that needs to be broken into its component sugars glucose and galactose—in order to be absorbed by the small intestine. Most humans stop making the enzyme sometime after they have been weaned and before adulthood. Once the enzyme is no longer produced, people feasting on dairy products will have the lactose pass through their small intestines undigested. When the sugar hits the colon, gut bacteria feast on it, producing gas and causing flatulence, nausea, diarrhea, and bloating.

Most Asians, southern Europeans, Africans, native Australians, Pacific Islanders, and Native Americans stop producing lactase by the time they are adults. By and large the only populations who effectively fail to wean as adults are northern Europeans and certain nomadic tribes in Africa and Arabia such as the Tuareg, Fulbe, Beja, and Bedoin people who rely on cattle and milk for their economy and their nutrition.

The genetic trait for lactase persistence is thought to have developed sometime in the past five thousand to ten thousand years while humans were domesticating animals and exploiting their milk for nutrition. Because the geographic distribution for lactase persistence mimics that of dairy farming, scientists have speculated that the trait conferred a significant advantage for those who could digest milk. Perhaps the gene allowed nomads in the desert to maintain a better electrolyte balance in arid conditions. Northern European populations who were exposed to much less sun may have benefited from the high levels of calcium in milk.

Whatever the benefit milk drinking conferred, Joel Hirschhorn and colleagues from Cornell University confirmed that it did provide a significant benefit when they analyzed selective pressure and polymorphisms in the lactase gene. They found that the gene had undergone the strongest positive selective observed for any gene in the human genome.

In order to conduct the genetic analysis, the Cornell team needed genetic markers for lactase persistence. These markers were provided in 2002 when Leena Peltonen of the UCLA School of Medicine and the University of Helsinki announced that she and her team had found changes in individual nucleotides at two different locations in a region just in front of the lactase gene that determined whether a person could digest milk. These single nucleotide polymorphisms, or SNPs, are located at 13,910 bases and 22,018 bases in front of the lactase gene. Much like NASA's mission control engineers counting down to lift off, scientists refer to the location of DNA bases in the promoter regions leading up to genes with a minus sign.

The team studied nine extended Finnish families as well as some Germans, Italians, and South Koreans. They found that when a person has a thymine in the -13,910 position on either copy of the lactase gene, they continue to digest milk into adulthood. However, if they inherit two copies of the gene with a cytosine at that position, they will become lactose intolerant. That one change, from a cytosine to a thymine, was seen in all of the people who were able to digest milk.

For the most part, people who carried a guanine at the second SNP found at -22018 bases lost the ability to digest lactose, while those who had an adenine in the position experienced lactase persistence. Peltonen and her colleagues found the -22018 allele showed some variability in effect.

These SNPs occur in the area in front of the lactase gene called a promoter, which serves to call the cellular machinery needed to transcribe the DNA into RNA. The lactase persistence is likely to result from a change in the regulatory element. Jesper Troelson at the University of Copenhagen demonstrated that a promoter containing the T-13910 SNP can activate lactase expression more vigorously than one with the C-13910 SNP in cell culture. His team is still working to identify how that happens and what transcriptional enhancers or repressors are at work.

While the SNP results are giving researchers new insight into the mechanism by which lactase persistence occurs, it has also provided a means to test for lactose intolerance. Current means for testing for lactose intolerance involve either consuming high levels of lactose after fasting and monitoring blood glucose levels or by measuring the amount of hydrogen produced by gut bacteria feasting on the undigested lactose. Neither test is particularly pleasant or effective.

The Finnish company Medix Laboratory was the first to offer the testing for Finns, but such testing based on the C/T-13910 allele may not prove applicable for everyone. Dallas Swallow at the University College London examined twenty different African populations, some of whom were milk drinkers and others who were not. Her team found the frequency of the persistence allele, T-13910, was too low to explain lactase persistence among Africans. As a result, she argues that the SNP cannot be the causative mutation for lactase persistence and calls for more extensive genotyping.

Because lactase persistence so closely matches dairy production, it may be possible to use the trait to follow human migration patterns, and that's just what Peltonen has attempted to do. By examining the lactase gene in thirty-seven populations on four separate continents, she searched for the origin of lactase persistence. She found that those groups that lived in region between the Ural Mountains and the Volga River, including the Udmurts, Mokshas, and Ezras, among others, were probably the first to develop lactase persistence approximately six thousand years ago. After this time, they spread the gene variation to Europe and the Middle East.

Wherever it came from, the persistence of lactase has proved a conundrum. Slowly but surely, the genetic story of lactase persistence is beginning to unravel, and soon it will tell us who's got milk and who doesn't.

Innate Sensing Genes

In 1990, Muppets creator Jim Henson, like millions of other Americans, felt the burning pain in the back of his throat that heralds the onset of a strep infection. Unlike those untold millions of fellow sufferers, Henson died from his battle with the bacteria known as streptococcus.

Henson, a robust man in his fifties, fell victim to a medical "perfect storm." While the Group A streptococcus bacteria are best known for causing strep throat, Henson contracted a particularly aggressive variety of the bacteria, which resulted in pneumonia. To make matters worse, his body responded with an enormous inflammatory response, causing his blood pressure to plummet as he went into shock. Henson developed sepsis, a dreaded and difficult-to-manage self-annihilative reaction to the bacteria in his bloodstream.

Henson's tragic death shocked the nation. How could a seemingly healthy middle-aged man die from such a common infection? To be certain, contracting an aggressive strain of Group A strep associated with both flesh-eating disease and sepsis didn't help. But simply having the bad luck to run across an especially virulent bug isn't enough to explain why the famed Muppeteer died. Most people who are infected with this type of bacteria live to tell the tale. Henson's immune system effectively went haywire, attacking not only the invading bacteria but his internal organs as well.

Despite living in the age of antibiotics, approximately two hundred thousand Americans die each year as a result of sepsis. To put that number into perspective, every year sepsis kills as many Americans as breast cancer and lung cancer combined. Even though it is triggered by bacterial infections that seep into the bloodstream, sepsis itself is not an infection. In fact, rather than curing the condition, antibiotics often exacerbate the medical crisis. Sepsis takes hold when our first line of defense against microbial invaders spirals out of control.

Animals possessing both a jaw and a spine depend on two very different immunological strategies to fight the microbial horde. The first identifies an infection and unleashes a chemical cascade that causes inflammation and flat-out kills the invading microbes. The second strategy involves identifying specific microbes and developing a long-lasting targeted response that remembers those microbes and stifles them before they can mount an infection. The first strategy is encompassed by the innate immune system, which is a generalized response to infections. The more specialized response is the domain of the adaptive immune system.

Sitting at the nexus of these two different immune strategies are a set of proteins known as the toll-like receptors. Essential to the workings of both immune systems, the toll-like receptors (TLRs) are the body's sentinels that identify microbial components and facilitate the communication between the two immune systems. Understanding the role these sentinels play may offer avenues to preventing the diseases caused when these two immune systems overreact.

TLRs came to light in the mid-1990s because they resembled a fruit fly protein that not only played a critical role in helping the developing fly determine back from front but also protected the fly from fungal infections. In 1997, Charles A. Janeway Jr., a Howard Hughes Medical Institute investigator at Yale University, discovered a human protein strikingly similar to that of the fly protein. Housed on cells that comprise the innate immune system, the ten human TLRs unearthed so far don't play any known role in development; instead, they trigger the activity of the innate immune system.

In 1998, Bruce Beutler, a Howard Hughes Medical Institute investigator now at The Scripps Research Institute in San Diego, and his colleagues at the University of Texas Southwestern Medical Center established TLRs as the immune system's sentinels by studying mice markedly incapable of responding to a bacterial molecule known as endotoxin or lipopolysaccharide (LPS). A molecular component of the "Gram negative" class of bacteria, LPS is usually a potent trigger of the innate immune system that causes fever and, in sufficient concentrations, death. Beutler's group found that LPS binds to TLR4 in order to trigger the innate immune system to initiate acute inflammatory responses by ramping up the expression of antimicrobial genes and inflammatory signaling molecules.

That finding led to a flurry of activity, which ultimately identified the microbial molecules that bind to the various members of the TLR family. These molecules, often referred to as pathogen associated molecular patterns or PAMPs, include conserved components of RNA and DNA viruses, "Gram positive" and "Gram negative" bacteria, fungi, and protozoa. Each of the TLRs binds to certain microbial PAMPs, but researchers haven't identified a PAMP that binds to TLR10. Sometimes two TLRs team up to bind to certain molecules, thereby increasing the TLRs' repertoire. In general, human TLRs 1, 2, 4, 5, and 6 recognize mainly bacterial products, and TLRs 3, 7, 8, and 9 specialize in detecting viral components.

While binding to these PAMPs is the first order of business for TLRs, in order to alert the immune system of a microbial threat, the binding must set off a signaling pathway, which triggers inflammation. TLRs accomplish this by associating with each other and other different molecules using a special amino acid sequence. Each TLR is made up of a section inside the cell compartment, a section traversing the cell membrane, and, finally, a section outside the cell. The section inside the cell carries a common pattern of protein building blocks known as the "Toll/interleukin 1 receptor/resistance" motif (TIR). This pattern of amino acids appears in plants as well as animals and usually heralds an innate immune function for both plants and animals. TLRs bind to the next molecules in the signaling pathway via these TIR motifs.

Only four proteins found inside the cellular compartment carry this motif. It is these four proteins that bind to the various TLRs and propagate the signal eventually activating hundreds of genes. While individual TLRs activate different genes and signal different aspects of the innate immune system, all TLRs detect microbial infections and trigger inflammation. Researchers are beginning to tease apart how those signals trigger the whole repertoire of innate immune response.

The innate immune system's inflammatory response is the first salvo in fighting off new infection. In order to create a memory of microbial interlopers, the adaptive immune system must be activated. The first step in that process begins with dendritic cells: immune cells that reside in areas of the body in contact with the environment (such as the skin and the gut) and capture antigens, which are the unique microbial components. Dendritic cells engulf the antigens, activate, and move to the lymph nodes, following signals initiated by TLRs.

Once dendritic cells are on the move, they undergo a maturing process that allows them to stimulate immature or naïve T cells by presenting T cells with an antigenic load. TLRs housed in dendritic cell walls directly trigger maturation of those cells by increasing the expression of genes that produce co-stimulatory molecules. Dendritic cells respond only to microbial threats for which they possess an appropriate TLR. In 2002, a team of researchers at Kansai Medical University, led by Ryuichi Amakawa, discovered that binding to TLR7 would induce the production of different proteins depending on what type of dendritic cell carried the TLR.

Not only do TLRs trigger the innate immune system cascade, they also control an important step in the immune cells that engulf microbial threats and dead host cells. Dendritic cells, macrophages, and neutrophils all demonstrate the ability to engulf other cells—a process known as phagocytosis. These immune cells first surround a bacterial threat with cell membrane and create a sac or vesicle (a phagosome), which encloses the threat. After the microbe has been internalized into a phagocytic cell, the phagosome fuses with a cell body (a lysosome) and creates a mature phagosome. It is in this mature phagosome where cellular enzymes destroy the invader. Innate immune system cells such as neutrophils and macrophages simply kill the invader and dump the by-products of the destruction. Dendritic cells kill the microbe and display the by-products so T cells can use the information to remember the invader.

In 2004, Ruslan Medzhitov, a Howard Hughes Medical Institute investigator at Yale University School of Medicine, discovered that TLRs directly and rapidly signal the phagosomes to mature. But TLRs demonstrated the ability only when the cells engulfed microbes, never when the phagocytic cell picked up dead host cells. The results indicate that thwarting this rapid maturation when a host cell is the target may provide some control over out-of-control immune responses, such as autoimmunity.

As signaling pathways become clearer, strategies for circumventing the hyperactive immune response that causes sepsis begin to emerge. In 2004 Carsten Kirschning and colleagues from the Technical University of Munich tested antibodies against TLR2 as a means for preventing shock following a lethal dose of the Gram positive bacteria cell wall components that TLR2 recognizes. The antibodies effectively swamped the TLR2 receptors, quenched the signal from macrophages, and disrupted the synthesis of inflammatory cytokines. The TLR2 "blockade" prevented mice from developing septic shock.

While the results hold promise for interrupting the path to sepsis, there remain several potential pitfalls. First, a number of therapies have attempted to interrupt the sepsis cascade and have shown promise in mice, only to fail stupendously when tested in humans. Second, the German group focused on obstructing the very first step in the inflammatory process. Patients at risk of, or suffering from, sepsis will most likely come to their physicians' attention far too late in the disease process for efforts at TLR2 blockade to prevent shock.

Still, understanding how these sentinels of the immune system coordinate the activities of the two arms of our immune systems to set into motion the intricate and interrelated events of our response to infectious threats provides hope that we may find new strategies for conquering both autoimmunity and sepsis.

The Sidedness Genes

As I comfort my fussy newborn daughter, I instinctively raise her tiny body to my left shoulder, shushing quietly into her ear. I hold her, swaddled close to me, mimicking the tight quarters of my womb, which had so recently been her home. Her chest to my chest, she feels the reassuring rhythm of my heart, which sounded as an ever-present metronome during the forty weeks she was growing inside me.

I am repeating the motions mothers have used for millennia to calm their babies. Still, I can't help having questions about this seemingly innate way of nurturing a newborn. Why do I choose to hold her to my left side? Is it simply because I am an obligate right-hander and want to pat her soothingly with my free hand? Or do I somehow sense the inherent asymmetry of my own body? As much as external symmetry is the norm for human beauty, the organs of our body don't simply line up at midline: the heart and spleen are on the left; the liver and gall bladder squeeze in to the right. Nearly all organs in the chest and abdomen adhere to a consistent left–right asymmetry. Even our two lungs have an asymmetric bearing—we have a two-lobed lung on the left and a three lobed lung on the right.

At least most of us do, that is. For one out of every eighty-five hundred people, it's as if they're peering through from the other side of a mirror: what should be right is left and vice versa. The condition is called situs inversus, and aside from confusing some green medical students and residents, it usually causes no problem. Curiously, the real trouble begins when the inversion isn't complete. Minor aberrations in left–right asymmetry can scramble the internal organs to such a degree that infants born with only a partial left–right inversion need emergency surgery as soon as they make their entrance into the world.

Minor or major, these left–right aberrations lead to some interesting developmental questions. Why, for example, is our internal asymmetry so consistent? If the positions of our viscera were simply the result of random events in development, we could expect half of us to have our hearts on the right, and the rest on the left. But we don't, and that fact has mystified developmental biologists for decades.

On a practical level, our internal asymmetry is a simple packing problem—there is a whole lot of stuff that needs to fit in a rather confined space, and compromises need to be made. For example, in order to fit the heart into the chest, the left lung has two lobes where the right has three, and the major airways into the lungs are tilted at different angles to free up even more space.

Spatial issues aside, a consistent asymmetry permits our organs to be plumbed together properly. It's easy to see how a complete inversion of the asymmetry would allow the organs to attach to one another in the right order (and no, people with situs inversus are no more likely to be left-handed than anyone else.) The concept of all-or-nothing rules the day, however. Partial body plan inversions result in having a mix of organs in the normal and inverted orientation, a situation referred to as heterotaxia, which makes it very difficult to link the organs together properly. The right ventricle of the heart needs to send blood to the lungs to pick up oxygen, and the left side needs to receive the blood from the lungs and send it out to the rest of the body. When those connections are scrambled, the body can't provide adequate oxygen for vital organs like the brain.

The most serious scenario arises when the organs are completely symmetrical down the midline of the body. The condition is called isomerism. Without an established sidedness, the blood vessels connecting other organs to the heart randomly attach themselves to the nearest available blood vessel. Such willy-nilly vascularization lands newborns immediately in the surgical unit in a desperate effort to save their lives.

Clinical complications aside, when an embryo begins to develop it has to somehow get its bearing in space. At first just a mass of dividing, identical cells, an embryo begins to develop axes lines that run head to tail and belly to back, marking which end is the head and which side is the front. How an embryo marks these two directions is fairly well understood.

The left–right decision is a far trickier matter, however. In order to set up left–right asymmetry, an embryo must break its bilateral symmetry in a consistently handed manner, setting up a new axis exactly perpendicular to the other two. Some have argued that an embryo first understands left from right when an asymmetrically shaped molecule lines up along the back–front and head–tail axes. If you imagine the letter *F* plastered to your chest, as long as it is positioned consistently with regard to the other two axes, the arms of the *F* will distinguish between left and right. The problem is that no such molecule has been described.

Since the 1930s, physicians have documented patients who suffer from sinus and respiratory problems, and coincidentally, also have situs inversus. It wasn't until 1976 that a theory about how the body achieved asymmetry was broached. Swedish physician Björn Afzelius was studying men with Kartagener's syndrome. All of the patients not only showed some problems with left–right asymmetry but also suffered from infertility. When Afzelius looked at the men's sperm, he discovered they couldn't move. The sperm tails, or flagella, were paralyzed. Flagella are just supersized versions of hairlike projections, which on other cells are called cilia.

In addition to the infertility complaints, the men also suffered from frequent respiratory infections. Afzelius looked at bronchial tissue from one of the men and noted that the cilia lining the bronchial tubes were also paralyzed because they lacked structures called dynein arms. Afzelius suggested a connection between the symptoms: perhaps when the cilia waved, they forced the developing organ to one side or another in the body. Like many scientific theories, this one was shelved for a couple of decades and later resurrected.

Nobutaka Hirokawa and his colleagues at Tokyo University breathed new life into the theory when they were studying a group of proteins called kinesins. The sole purpose of these proteins is to haul molecular cargo along the cell's internal scaffolding, or cytoskeleton. Having identified a new kinesin complex, the group decided to engineer mice lacking the gene for one of the components of the complex—the dynein heavy chain—in order to determine exactly what it did.

Whatever that component did, it was critical. All of the mice died as early embryos, and approximately half had reversed left–right patterning. The Japanese researchers took a special look at an embryonic structure called the node, which was known to have a role in left–right patterning. A triangular patch of cells that forms a pit at the head end of the developing embryo, the node establishes the head-to-tail axis as well as the overall body plan. In mouse embryos, the node is normally covered in cilia; the node in Hirokawa's mouse embryos had no cilia at all.

Hirokawa's group decided to examine normal mouse embryos under an electron microscope and found the cilia gently swirling in a counterclockwise manner. When they added buoyant, fluorescent beads to the liquid used to study the nodal cells, they discovered that the cilia generated a left-leaning current or nodal flow. The researchers suggested the nodal flow triggered left–right asymmetry by allowing chemical signals that activate genes needed during development to waft over to one side of the embryo. As a result, specific genes were activated on the left or the right side of the embryo.

Hirokawa furthered this theory in an experiment in another mouse strain called *inv*, for inverse. Approximately 85 percent of the newborn rodents in this strain display situs inversus. The researchers found the nodal cilia in these mice was motile but could produce only very weak leftward nodal flow.

While suggestive of the importance of nodal flow in the development of left–right asymmetry, these experiments didn't expressly show that the direction of the flow altered symmetry in any way. In a 2002 experiment that Harvard University developmental biologist Cliff Tabin calls a "technical tour de force," Shigenori Nonaka, Hiroshi Hamada, and colleagues at Osaka University in Japan mechanically reversed the nodal flow in mouse embryos. In normal mouse embryos, reversing nodal flow also reversed the embryo's visceral left–right orientation. In other words, these embryos develop a complete inversion. Mechanically reversing nodal flow in embryos lacking cilia all together restored a consistent left–right asymmetry.

Hirokawa's model was proving an elegant explanation for left–right asymmetry. And, scientists love nothing more than an elegant biological model backed up by genetic evidence. The problem with the hypothesis from the outset, however, has been proving that nodal flow was the very first step in the development of sidedness.

As yet, no other gene or molecule displays a preference for side prior to the development of the nodal cilia in mouse embryos. Shortly after the publication of Hirokawa's nodal-flow model, evidence from frog and chick embryos indicated the nodal flow wasn't the first step in establishing sidedness in those species. In chick and frog embryos, several genes were active on one side or the other *prior* to the development of cilia on the node.

At least for chicks and frogs, a nodal flow generated by cilia appeared to be an unlikely source for that initial development of sidedness. Michael Levin, now at the Forsyth Institute in Boston, and Mark Mercola, at Harvard University Medical School, examined the portals between cells in chick and frog embryos. These cellular portals, also known as gap junctions, are specialized regions of cell membranes that physically connect neighboring cells and provide these cells a means to exchange small molecules such as calcium and potassium ions as well as amino acids and nucleotides. Levin and Mercola discovered the activity across the gap junctions displayed sidedness prior to the development of nodal cilia in both chick and frog embryos. Additionally, the gap junction asymmetry was required for the development of sidedness—without it, the embryo ultimately failed to distinguish left from right.

Even gap junction asymmetry isn't the first molecular indication of sidedness in frog embryos. Levin and Mercola identified a cell-membrane protein called H+/K+-ATPase that concentrates its activity on the right side of the embryo a full day before the cilia show up in the frog embryo. In addition to H+/K+-ATPase, Joseph Yost and his colleagues at the University of Utah in Salt Lake City reported the protein syndecan-2 is active only on the right side of the frog embryo before cilia show up. Moreover, interfering with the expression of either of these proteins obstructs normal left–right patterning in frog embryos.

With evidence mounting that gene expression and other cellular processes in some vertebrate embryos experience sidedness before cilia appears, questions are brewing about whether the appearance of cilia is truly the first step in left–right patterning.

Utah's Yost believes "there is strong evidence for cilia participating in asymmetry." Levin agrees, but both researchers assert nodal flow isn't the step that initiates symmetry in the first place. The nodal-flow hypothesis relies on cilia setting up a current that unevenly distributes a signaling molecule throughout the embryo to establish sidedness. The problem with this model, according to Levin, is that none of these proposed signaling molecules has been identified.

Even so, the genetic evidence seems to pinpoint functioning cilia as the source of left–right asymmetry. After all, humans and mice that lack functioning cilia develop severe aberrations in left–right patterning. Yost points out that both humans and mice that lack the dynein protein have many more problems than reversed visceral orientation. Dyneins play critical roles in a number of cellular transport processes, so it is possible that problems with nodal cilia formation are ancillary to the defects that actually cause patterning problems. In other words, the defective cilia could be indicators rather than causes of the failure to develop sidedness.

Because gap junctions and ion pumps show sidedness before the development of cilia in the node, Levin has proposed that electric fields may play a role in establishing left from right. Still, he readily admits even if gap junctions ferry signaling molecules that establish sidedness, he doesn't know what initially breaks symmetry either.

While eminent scientists develop elegant theories one day only to begin tearing them down the next, I will still raise my daughter to my left shoulder and heart-to-heart comfort her for as long as she'll let me, confident in the knowledge that be it the result of cilia or ions, my heart is in the right place.

Epilogue

Snippets of Information Reveal the Whole?

From the early work of Archibald Garrod to the present understanding of Huntington's disease, single human genes displaying frank Mendelian inheritance patterns and stark phenotypes provide for fascinating stories; they have been the principle means by which science has gained insight into how genetics works in the human body. Nevertheless, the most profound understanding of human genetics won't come as a result of these single gene discoveries. Most human diseases and conditions arise from the complex interplay of multiple genes, behavior, and environmental exposures. As we've seen, scientists have discovered a number of genes amplifying the risk for developing common human diseases such as diabetes, cardiovascular disease, and various forms of mental illness. When dealing with these complex disorders, however, any single gene represents only a tiny percentage of the variability of the disease. As a result, identifying the genes involved in diseases like diabetes and heart disease as well as conditions such as obesity has been tremendously slow going. Until now, that is.

With the Human Genome Project providing the sequence of nearly all of the 3 billion DNA base pairs that make up human chromosomes, scientists have moved forward to catalog the little ways in which those 3 billion base pairs differ from person to person. It's important to remember that 99.9 percent of human DNA is similar, even between unrelated people—only about one in every twelve hundred DNA bases differs between people. The variation in that last 0.1 percent of human DNA results in visible differences such as hair and eye color and susceptibility to diseases, and the bulk of that variation comes in the form of a single DNA base-pair change. These are called single nucleotide polymorphisms (SNPs—pronounced snips).

Using the tools of biotechnology, researchers have created two maps of SNPs that will serve to link patterns of SNP variation to human traits: a private effort by California-based Perlegen Sciences Inc., and the International HapMap Project, a worldwide public effort that includes scientists from Canada, China, Japan, Nigeria, the United Kingdom, and the United States. In February 2005, Perlegen provided a first draft of its SNP map comparing seventy-one Americans of European, African, and Asian descent. The HapMap Project reached completion in October 2005 when it cataloged a denser map of SNPs in 270 individuals from China, Japan, Nigeria, and Americans of European descent. By using these maps to discern patterns associated with disease and response to drugs, researchers hope to finally move into the era of personalized medicine where physicians will prescribe therapies and pharmaceuticals based on a patient's likelihood of benefiting from such prescriptions.

The goal of the SNP maps isn't to identify disease-causing genes directly. Instead, the maps will provide groups of SNPs close to each other on the genome that tend to be inherited together. These groups are called haplotypes. With a haplotype map in hand, researchers can start asking detailed questions about genotype and phenotype. For example, does a certain haplotype pattern appear in people suffering from a complex disease like cardiovascular disease? If the answer is yes, researchers would know that genes in some way associated with the disease are in or near the sections of DNA making up those haplotypes.

"For so long we have heard talk of the promise of genetics," noted David Cox, chief scientific officer and cofounder of Perlegen, at the annual meeting of the American Association for the Advancement of Science in February 2005, where his company announced it had genotyped 1.6 million SNPs. "[The map] is a powerful new resource for realizing this promise. We've started the clock and we should see progress in the next couple of years."

One of the most frustrating aspects of drug development is finding a drug that works great for some people and not at all for others. Even more worrisome are cases in which a drug causes serious side effects in some people but not in others. For example, the irritable bowel syndrome (IBS) drug Lotrenox® (alosetron hydrochloride) was originally approved for the American market in early 2000 as a treatment for women with diarrhea-associated IBS. For some women the drug was a blessing that allowed them to live a more comfortable, normal life. For others, Lotrenox® triggered a life-threatening side effect called ischemic colitis, a situation in which blood can't reach the intestines. In November 2000, GlaxoSmithKline voluntarily pulled the drug from the market as a result of this severe side effect. However, many women who benefited from the drug were dismayed that the only drug able to relieve their symptoms was no longer available. After careful study, the FDA agreed to allow the company to market the drug on the condition that the women who were prescribed it were closely monitored, and physicians prescribing it participated in the company's monitoring program. What accounts for some women enjoying relief from IBS symptoms and others suffering severe consequences remains unknown.

Most scientists and physicians, however, suspect that individual SNP patterns are the source of variability in drug performance; except in rare instances, it has been almost impossible to figure out who the best candidates are for particular therapies.

Comparing the SNP patterns of people who respond best to a therapy with those who suffer side effects will identify the SNPs responsible for the differences. Even without knowing which genes trigger side effects, physicians could use a simple blood test to prescribe medicine to those most likely to benefit and prevent those most likely to have bad side effects from taking the medicine in the first place.

"[The SNP map] will allow drug companies to take treatments that don't work all that well and make them better," Cox says. "This resource can be a tool to inform the treatment of individual patients. It will allow scientists to use genetic information in a practical way . . . while I'm still alive."

The wealth of information the SNP maps contain will provide new insight into biology and medicine, but they may also cause some ethical dilemmas. Troy Duster, director of the Institute for the History of the Production of Knowledge at New York University, is concerned that with the SNP maps in hand, some will try to discern the genetic haplotype of race.

"The goal of the project(s) is admirable," Duster says. "However, I expressed concerns about how the data is being reported out. One of the convenient ways is to use ancestral heritage and race is a proxy for that."

Duster is concerned that the information from the SNP maps will be used to reinscribe racial taxonomies across a broad range of scientific practices and fields. That becomes a problem for a number of reasons, not the least of which is the fact that there are no discernable genetic boundaries for race. Cox agrees that the variation in haplotypes doesn't form stark lines delineating one race from another but rather a gradient from one end of the earth to another. As a result, Duster notes that race is a sociological concept, not a biological one.

"What I would recommend is that any time geneticists report population data, they should always attach a caveat," Duster says. "It could read something like this, 'allelic frequencies vary between any selected human groups—to assume those variations reflect racial categories is unwarranted.'"

While the SNP maps can't delineate social constructs like race, there will also be medical variation that can't be explained by genetics. David Altshuler of the Broad Institute of Harvard and the Massachusetts Institute of Technology in Cambridge,

Massachusetts, and participant in the International HapMap project, notes that even with human genetic variation cataloged, genetics won't explain all of the variation seen in drug response. He points out, "The (SNP) work is clearly exciting, but [SNPs] aren't the only thing involved. The act of drinking grapefruit juice alters the body's ability to clear certain drug substances."

In fact, the antirejection drug cyclosporine, which is routinely given to patients who've undergone organ transplantation, is affected by grapefruit juice. The drug isn't metabolized as quickly by the body when a patient drinks the juice. As a result, the drug's effects last longer or are stronger. Some patients have even used grapefruit juice as a means to reduce the amount of drug they needed to buy and take. Other such environmental effects most certainly exist, and no genetic tool will explain differences that result from them.

While the SNP maps won't provide all the definitive answers to medical and biological questions, they do open the door to a new world where complex genetic interactions and minute effects can be identified. And with that, our understanding of what genes do and what they don't do will continue to grow.

Appendix

A Genetics Primer

If you're reading this section, you've taken my advice and decided to take a brief genetics refresher course. The field of genetics, like any scientific pursuit, uses some arcane terms and makes shorthand references to rather complicated processes. Here is your opportunity to brush away the cobwebs clinging to the information gathered from your last biology course and get a handle on the biological intricacies of the human genome. You'll find important concepts highlighted in bold. And, as I stated in the introduction, I promise this will be painless.

It's probably helpful to zoom in on human genes starting from the outside. Just take a look at anyone you know. Do they have dark eyes? Dark skin? Their mother's mouth? Their father's nose? Perhaps they're albino. Maybe they're exceptionally short or tall. All of these characteristics or conditions are called **traits,** or in genetic parlance, **phenotypes**. With the exception of some artful work by plastic surgeons, they are genetic in origin. Other less visible things like blood type, ability to clear cholesterol from the body, and ability to metabolize certain amino acids are also traits. That's not to say all traits are inherited. Some traits are acquired, such as the ability to throw a curveball, play the piano, or land the perfect triple Lutz. But for now, we'll concern ourselves solely with those traits that have some heritable component.

To get a better understanding of how those traits come into being, we have to move in a little closer to the level of a cell. It doesn't matter if you're talking about a nerve cell, a skin cell, or a brain cell; with the exception of a mature red blood cell, they all carry a complete complement of all human genes. Cell structure is such that genetic material resides in an inner compartment called the **nucleus**. Genetic material is housed on **chromosomes**. Chromosomes get their name from the Greek for color and body; when the cell is stained and placed under a microscope, chromosomes show up as darkly stained thread-like bodies. In humans, all cells, except red blood cells which have no nuclei, house all forty-six chromosomes.

In addition to sucking up certain dyes, chromosomes have other characteristic features. First, each chromosome is an amalgamation of a long contiguous stretch of **DNA** (deoxyribonucleic acid, the molecule that stores genetic information) and associated proteins. When cells are doing the work they normally do, their chromosomes exist as superfine tangles. But when the cell is in the process of duplicating and dividing, the chromosomes become visible as thin rods with a "waist," and this indentation is known as the **centromere**. While serving as an anchor point for separating duplicated chromosomes during cell division, the centromere also makes a terrific reference point for scientists who are trying to pinpoint certain genes. Because the chromosomal indentation isn't directly in the center of the chromosomes, scientists refer to the **short arm** and the **long arm** of a particular chromosome when they're homing in on a specific gene.

The human genome contains twenty-two pairs of chromosomes plus the two sex chromosomes: In females, that would be two X chromosomes; in males, sex chromosomes include an X and a Y chromosome. The non-sex chromosomes are referred to as **autosomes** to distinguish their inheritance patterns from the sex chromosomes. The pairing results from each person receiving one full set of chromosomes from the mother and another full set from the father. As for the sex chromosomes, everyone receives an X chromosome from mom and either an X or Y chromosome from dad.

Simply defined, a **genome** is all of the genes available to an organism whether it be a human being, dog, or bacteria. A more complete definition would also include the vast stretches of DNA that clearly aren't genes but have an as-yet-undefined role in biology. This so-called "junk DNA" or "selfish DNA" comprises as much as 97 percent of the DNA in the human genome. Research by Adam Siepel and David Haussler, a Howard Hughes Medical Institute investigator at the University of California, Santa Cruz, shows that the proportion of this junk DNA conserved between species increases as you move up the evolutionary tree from yeast to insects to vertebrate animals. Unless Mother Nature is an incurable packrat, it doesn't seem likely this junk DNA is hanging around because it's useless. So, for our purposes now, the genome is all of the DNA comprising the chromosomes.

In a world where bigger is better, a bigger genome doesn't necessarily translate into a more complex organism. For example, the human genome contains roughly twenty-five thousand (about one-fourth of the genes that researchers speculated they would find when the Human Genome Project was initiated in the early 1990s) while the rice genome comprises approximately sixty thousand genes.

Moving in a little closer, we can start to look at **genes**. Genes reside in sections of DNA comprising the chromosome. For all the focus on genes and disease, the genes themselves don't do much. Genes are really just an instruction set or blueprint to tell the cell how to make the real movers and shakers: **proteins** and **RNAs** (ribonucleic acids). When genes have subtle differences between them that cause different traits, they're called **alleles**. A single gene can have multiple alleles. For example, ABO blood type represents three alleles, A, B, and O. Which alleles are represented in your genome is referred to as your genotype. In rare instances, traits can be linked to single genes; for example, achondroplasia, a form of dwarfism resulting in short limbs and characteristic facial features, occurs when a single gene is mutated. Usually, traits like weight and height are the reflection of the interplay among a number of genes and environmental factors like pre- and postnatal exposure to chemicals, smoking, and nutrition.

In order to understand what genes actually are, we need to move in a little closer to the level of DNA and its architecture. Chromosomes aren't simply a single strand of DNA snaked around associated proteins like **histones**—proteins found in the nucleus that act like a spool for DNA. In order for DNA to be stable enough to serve as a reference every time the cell needs to make a new protein, it exists as two intertwined strands: the oft-mentioned "**double helix**." DNA is usually described as a backbone of alternating sugar and phosphate atoms with one of four **bases** projecting from the sugar molecule. The larger of the bases, **purines**, are adenine and guanine. Cytosine and thymine, or **pyrimidines**, are the other two bases. When scientists refer to these bases, they usually do it by their first initials: A, T, G, and C. A key feature of the double helix is that the two strands that make it up are **complementary**. There is always a large purine across from a smaller pyrimidine and a small pyrimidine across from a larger purine. Additionally, the bases have specific partners: If one strand has an A at a certain position, the

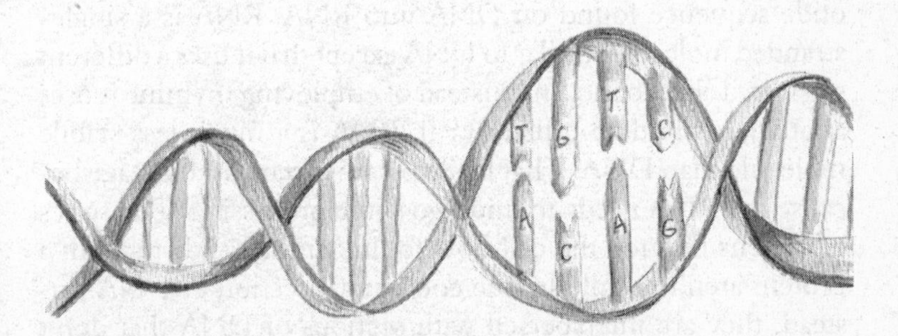

The double helix is a molecule of two complementary DNA strands. The nucleotide bases—adenine and thymine, guanine and cytosine—always pair with each other through weak chemical bonds. Those pairings serve as the basis for the genetic code.

other strand always has a T and vice versa. Likewise, if one strand has a G on it, the other strand has C and vice versa. The human genome has about 3 billion of these A, C, T, and G bases strung together to form the forty-six different human chromosomes.

Identifying the sequence of all the bases in the human genome is a massive achievement, but in doing so one is far from being able to read the biological instructions stored therein. If one were to look at the human genome as a book describing human life, then the nucleotide bases are all the letters in the tome, and genes are the sentences. In order to read the information in the human genome, one must know not only all the letters but the words and punctuation as well. The words in this instance are **codons**—three nucleotide sequences such as GAA or CTG. With four bases to choose from for each of the three spots in a codon, a little bit of math provides for sixty-four possible codons. As it turns out, with three exceptions, these codons represent one of twenty amino acids used to make proteins. So in the end, a gene is a series of codons that tell a cell how to build a protein by linking together amino acids.

In order to bring those words to life, the cell goes through a two step process. First, the gene must be **expressed**. One of the most important things to remember is that DNA doesn't do anything much by itself—it's just a blueprint. Therefore, an entire set of specialized proteins exists to transcribe the nucleotide sequence found on DNA into RNA. RNA is a single-stranded molecule similar to DNA except that it uses a different sugar in its backbone, and instead of employing thymine it uses another pyrimidine called uracil. RNA is a much less stabile molecule than DNA. That lability is important at this stage because the RNA needs to undergo some processing. The series of codons that tell the cell how to link amino acids to form a protein aren't usually in one contiguous section of DNA. Instead, they are interspersed with sections of DNA that don't code for the protein. The coding bits in the RNA transcript are called **exons**, and the noncoding bits are called **introns**.

In order to make a functional protein, the introns must be spliced out of the transcript and the exons linked to one an-

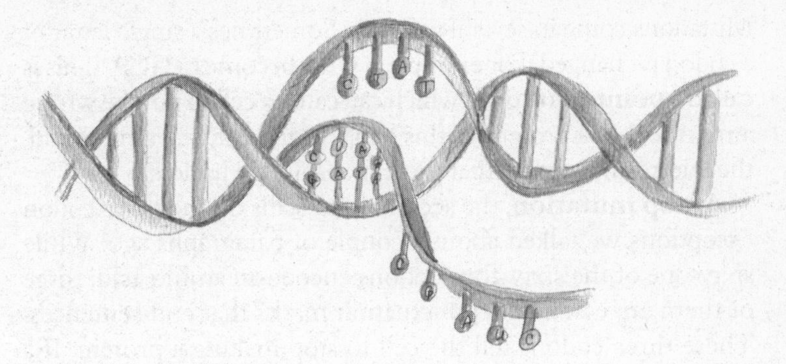

DNA serves as the template for cellular proteins to make a single-stranded molecule called RNA. The bases match adenine to uracil, thymine to adenine, cytosine to guanine, and guanine to cytosine. RNA makes a sort of nucleic acid negative, which is then translated into protein.

other. Once an edited RNA transcript is available, another set of cellular machinery uses the information on the RNA transcript to read the codons and assemble a protein using individual amino acids.

Once again, a gene isn't an active agent on its own, it's just the instruction set for making a specific protein, and proteins, with some notable exceptions, are the molecules that actually do the work in a cell like generating and using energy, responding to the environment, forming cellular components, and creating messaging systems that allow cells to communicate with each other. As a result, if there is a defect—known as a **mutation**—in a gene, the appropriate cellular machinery fails to assemble a specific protein or assembles it incorrectly, and it doesn't work very well. (Just a little aside: sometimes the end product of a gene is actually an RNA molecule. Mutations in such a gene will produce errors in the sequence of the RNA, which can result in nonfunctional or poorly functioning RNA molecules.)

Mutations occur when a series of codons misspell a gene. This can happen as a result of radiation, carcinogens, chemicals that damage DNA, or simply through a mistake sometimes made when a dividing cell makes another copy of its own DNA.

Mutations come in several varieties. Sometimes a single letter of a codon is changed. For example, a GAA becomes a GCA. This is called a **point mutation**, which can cause a cell to put the wrong amino acid in a protein. If this happens to be a vital amino acid, the end result renders that protein absolutely useless.

A **stop mutation**, the second kind, calls on the three codon exceptions we talked about a couple of paragraphs ago. While sixty-one of the sixty-four codons encode an amino acid, three of them are essentially punctuation marks that end sentences. These three codons tell the cell to stop making a protein. If a mutation in a gene causes a stop codon to show up prematurely, then the protein remains unfinished, and, it too is useless.

Genes also can be disrupted during normal cellular replication and division when chromosomes swap information between homologous sections. This recombination works to shuffle alleles—an important process when testes or ovaries are making sperm and eggs. At the same time, should anything go awry in the process of the swap, genes or parts of genes may be lost entirely.

For decades, mutations have been the lifeblood of genetics because the gene defects that occur allow a form of reverse engineering. By identifying the problems caused when mutations make a gene unable to produce usable protein, scientists can home in on the gene involved and learn how the protein encoded by that gene should normally function. Not surprisingly, scientists found it easiest to identify single genes that, when mutated, resulted in profound and sometimes fatal phenotypes such as mental retardation, clotting problems, and various devastating metabolic problems. Physicians and researchers readily tracked these diseases in affected families and established these diseases as genetic.

Unfortunately, because scientists identified and understood these types of genetic diseases first, many people developed a skewed view of the role of genes in biology. Too often nonscientists view genes and disease as looking for *the* gene that causes cancer or heart disease or obesity. No single gene embodies the complexity of these common human disorders. Instead, these mundane maladies arise from a complex interplay between many genes, environmental factors like exposure to radon gas, and behaviors such as smoking.

With the human genome laid bare, scientists are discovering how genes with only a small effect interact with other genes, the environment, and behaviors to cause diseases.

In addition, the human genome and the newest tools of genomics allow researchers to probe those sections of the genome that don't code for protein but affect the activity of genes by promoting the gene's expression by the cellular machinery. Each of these achievements provides important insights into the way genes influence the development and outcomes of complex diseases. These genetic discoveries will eventually alter the popular perception of genes as destiny.

References

Griffiths, Anthony J. F., Jeffery H. Miller, David T. Suzuki, Richard C. Lewontin, and William M. Gelbart. 1976, 1981, 1986, 1989, 1993, 1996, 2000. *An Introduction to Genetic Analysis.* New York: W. H. Freeman and Company.

Jorde, Lynn B., John C. Carey, Michael J. Bamshad, and Raymond L. White. 1995, 2000. *Medical Genetics.* St. Louis, MO: Mosby, Inc.

Mange, Elaine Johansen, and Arthur P. Mange. 1995, 1999. *Basic Human Genetics.* Sunderland, Massachusetts: Sinauer Associates Inc.

Online Mendelian Inheritance in Man, OMIM™. McKusick-Nathans Institute for Genetic Medicine, Johns Hopkins University (Baltimore, MD), and National Center for Biotechnology Information, National Library of Medicine (Bethesda, MD), 2000. http://www.ncbi.nlm.nih.gov/omim/

Introduction

Sturtevant, Alfred H. 1965, 1967, 2000. *A History of Genetics.* Cold Spring Harbor, New York: Cold Spring Harbor Laboratory Press and Electronic Scholarly Publishing Project.

When a Gene Makes You Smell Like a Fish

Cashman, J. R., K. Camp, S. S. Fakharzadeh, P. V. Fennessey, R. N. Hines, O. A. Mamer, S. C. Mitchell et al. 2003. "Biochemical and clinical aspects of the human flavin-containing monooxygenase form 3 (FMO3) related to trimethylaminuria." *Curr Drug Metab* 4 (2): 151–70.

Cashman, J. R. 2000. "Human flavin-containing monooxygenase: substrate specificity and role in drug metabolism." *Curr Drug Metab* 1 (2): 181–91.

———. 2004. "The implications of polymorphisms in mammalian flavin-containing monooxygenases in drug discovery and development." *Drug Discov Today* 9 (13): 574–81.

Christensen, Damaris. 1999. "What's That Smell: Modern Science puts its mark on a rare but ancient body-odor disease." *Science News*. May 15.

Yamazaki, H., M. Fujieda, M. Togashi, T. Saito, G. Preti, J. R. Cashman, and T. Kamataki. 2004. "Effects of the dietary supplements, activated charcoal and copper chlorophyllin, on urinary excretion of trimethylamine in Japanese trimethylaminuria patients." *Life Sci* 74 (22): 2739–47.

Zhang, J., Q. Tran, V. Lattard, and J. R. Cashman. 2003. "Deleterious mutations in the flavin-containing monooxygenase 3 (FMO3) gene causing trimethylaminuria." *Pharmacogenetics* 13 (8): 495–500.

Chapter 1: It Takes Two to Tango

Encyclopædia Britannica Online, s.v. "Mendel, Gregor." 2005. Encyclopædia Britannica Premium Service. http://www.britannica.com/eb/article-9051973.

The First Gene

Beadle, G. W. 1964. *Nobel Lectures, Physiology or Medicine*. 1942–62. Amsterdam: Elsevier Publishing Co.

Fernandez-Canon, J. M., B. Granadino, D. Beltran-Valero de Bernabe, M. Renedo, E. Fernandez-Ruiz, M. A. Penalva, and S. Rodriguez de Cordoba. 1996. "The molecular basis of alkaptonuria." *Nat Genet* 14 (1): 19–24.

Garrod, A. E. 2002. "The incidence of alkaptonuria: a study in chemical individuality. 1902 [classical article]. *Yale J Biol Med* 75 (4): 221–31.

La Du, B. N., and V. G. Zannoni. 1955. "The tyrosine oxidation system of liver. II. Oxidation of p-hydroxyphenylpyruvic acid to homogentisic acid." *J Biol Chem* 217 (2): 777–87.

Phenylketonuria and the First Genetic Test

Armstrong, M. D., and F. H. Tyler. 1955. "Studies on phenylketonuria. I. Restricted phenylalanine intake in phenylketonuria." *J Clin Invest* 34 (4): 565–80.

Bickel, H., and W. Gruter. 1957."Phenylketonuria with normal intelligence quotient; comparison of biochemistry & psychodiagnostic tests of 2 girls almost the same age" *Z Kinderheilkd* [In German] 79 (5): 509–21.

Kirkman, H. N. 1982. "Projections of a rebound in frequency of mental retardation from phenylketonuria." *Appl Res Ment Retard* 3 (3): 319–28

Koch, R., B. Burton, G. Hoganson, R. Peterson, W. Rhead, B. Rouse, R. Scott et al. 2002. "Phenylketonuria in adulthood: a collaborative study." *J Inherit Metab Dis* 25 (5): 333–46

Koch, R., E. Friedman, C. Azen, W. Hanley, H. Levy, R. Matalon, B. Rouse et al. 2000. "The International Collaborative Study of Maternal Phenylketonuria: status report 1998." *Eur J Pediatr* 159 (Suppl. 2): S156–60.

Levy, H. L. 1999. "Phenylketonuria: Old disease, new approach to treatment." *PNAS* 96 (5): 1811–13.

Lenke, R. R., and H. L. Levy. 1980. "Maternal phenylketonuria and hyperphenylalaninemia. An international survey of the outcome of untreated and treated pregnancies." *N Engl J Med* 303 (21): 1202–8.

Paul, D. 2000. "A double-edged sword." *Nature* 405:515.

Woolf, L. I. 1986. "The heterozygote advantage in phenylketonuria." *Am J Hum Genet* 38 (5): 773–75.

Maple Syrup Urine

Love-Gregory, L. D., J. Grasela, R. E. Hillman, and C. L. Phillips. 2002. "Evidence of common ancestry for the maple syrup urine disease (MSUD) Y438N allele in non-Mennonite MSUD patients." *Mol Genet Metab* 75 (1): 79–90.

Menkes, J. H., P. L. Hurst, and J. M. Craig. 1954. "A new syndrome: progressive familial infantile cerebral dysfunction associated with an unusual urinary substance." *Pediatrics* 14:462–66.

Morton, D. H., K. A. Strauss, D. L. Robinson, E. G. Puffenberger, and R. I. Kelley. 2002. "Diagnosis and treatment of maple syrup disease: a study of 36 patients." *Pediatrics* 109 (6): 999–1008.

Morton, D. H. 1994. "Through my window—remarks at the 125th year celebration of Children's Hospital of Boston." *Pediatrics* 94 (6): 785–91.

Puffenberger, E. G. 2003. "Genetic heritage of the Old Order Mennonites of southeastern Pennsylvania." *Am J Med Genet C Semin Med Genet* 121 (1): 18–31.

The Tangier Island Gene

Assmann. G, and A. M. Gotto Jr. 2004. "HDL cholesterol and protective factors in atherosclerosis." *Circulation* 109_(Suppl. 1): III-8,III-9,III-10.

Clee, S. M., A. H. Zwinderman, J. C. Engert, K. Y. Zwarts, H. O. Molhuizen, K. Roomp, J. W. Jukema et al. 2001. "Common genetic variation in ABCA1 is associated with altered lipoprotein levels and a modified risk for coronary artery disease." *Circulation* 103 (9):1198–205.

Fredrickson, D. S. 1964. "The Inheritance of High Density Lipoprotein Deficiency (Tangier Disease)." *J Clin Invest* 43 (2): 228–236.

Rust, S., M. Walter, H. Funke, A. von Eckardstein, P. Cullen, H. Y. Kroes, R. Hordijk, J. Geisel et al. 1998. "Assignment of Tangier disease to chromosome 9q31 by a graphical linkage exclusion strategy." *Nat Genet* 20 (1): 96–98. Erratum in: 1998. *Nat Genet* 20 (3): 312.

Rust, S, M. Rosier, H. Funke, J. Real, Z. Amoura, J. C. Piette, J. F. Deleuze, et al. 1999. "Tangier disease is caused by mutations in the gene encoding ATP-binding cassette transporter 1." *Nat Genet* 22:352–55.

Schaefer, E. J., C. B. Blum, R. I. Levy, L. L. Jenkins, P. Alaupovic, D. M. Foster, H. B. Brewer Jr. et al. 1978. "Metabolism of high-density lipoprotein apolipoproteins in Tangier disease." *N. Engl J Med* 299:905–10.

Singaraja, R. R., E. R. James, J. Crim, H. Visscher, A. Chatterjee, and M. R. Hayden. 2005. "Alternate transcripts expressed in response to diet reflect tissue-specific regulation of ABCA1." *J Lipid Res* 46 (10): 2061–71.

Timmins, J. M., J. Y. Lee, E. Boudyguina, K. D. Kluckman, L. R. Brunham, A. Mulya, A. K. Gebre et al. 2005. "Targeted inactivation of hepatic Abca1 causes profound hypoalphalipoproteinemia and kidney hypercatabolism of apoA-I." *J Clin Invest* 115 (5): 1333–42.

Wheatley, Harold G., and Harvey David Alan. 1973. "This is my island, Tangier." *National Geographic* 144 (5): 700–25.

The Celtic Curse

Asberg, A., K. Thorstensen, K. Hveem, and K. S. Bjerve. 2002. "Hereditary hemochromatosis: the clinical significance of the S65C mutation." *Genet Test* 6 (1): 59–62.

Beutler. E. 2005. "Hemochromatosis: Genetics and Pathophysiology." *Annu Rev Med.* Aug. 5. Epub ahead of print: http://arjournals.annualreviews.org.

Drakesmith, H., E. Sweetland, L. Schimanski, J. Edwards, D. Cowley, M. Ashraf, J. Bastin, and A. R. Townsend 2002. "The hemochromatosis protein HFE inhibits iron export from macrophages." *Proc Natl Acad Sci USA* 99 (24): 15602–7.

Fackelmann, Kathleen. 1997. "Rusty Organs: Researchers identify the gene for iron-overload disease." *Science News*, January 18.

Feder, J. N., A. Gnirke, W. Thomas, Z. Tsuchihashi, D. A. Ruddy, A. Basava, F. Dormishian et al. 1996. "A novel MHC class I-like gene is mutated in patients with hereditary haemochromatosis." *Nat Genet* 13 (4): 399–408.

Lee, P., H. Peng, T. Gelbart, and E. Beutler. 2004. "The IL-6- and lipopolysaccharide-induced transcription of hepcidin in HFE-, transferrin receptor 2-, and beta 2-microglobulin-deficient hepatocytes." *Proc Natl Acad Sci USA* 101 (25): 9263–65.

Livesey, K. J., V. L. Wimhurst, K. Carter, M. Worwood, E. Cadet, J. Rochette, A. G. Roberts et al. 2004. "The 16189 variant of mitochondrial DNA occurs more frequently in C282Y homozygotes with haemochromatosis than those without iron loading." *J Med Genet* 41 (1): 6–10.

Lucotte. G. 1998. "Celtic origin of the C282Y mutation of hemochromatosis." *Blood Cells Mol Dis.* 24 (4): 433–38.

Waalen, J., V. Felitti, T. Gelbart, N. J. Ho, and E. Beutler. 2002. "Penetrance of hemochromatosis." *Blood Cells Mol Dis* 29 (3): 418–32.

Metallic Madness

Bull, P. C., G. R. Thomas, J. M. Rommens, J. R. Forbes, and D. W. Cox. 1993. "The Wilson disease gene is a putative copper transporting P-type ATPase similar to the Menkes gene." *Nat Genet* 5 (4): 327–37. Erratum in: 1994. *Nat Genet* 6 (2): 214.

Petrukhin, K., S. G. Fischer, M. Pirastu, R. E. Tanzi, I. Chernov, M. Devoto, L. M. Brzustowicz et al. 1993. "Mapping, cloning and genetic characterization of the region containing the Wilson disease gene." *Nat Genet* 5:338–43

Petrukhin, K, S. Lutsenko, I. Chernov, B. M. Ross, J. H. Kaplan, and T. C. Gilliam. 1994. "Characterization of the Wilson disease gene encoding a P-type copper transporting ATPase: genomic organization, alternative splicing, and structure/function predictions." *Hum Mol Genet* 3 (9): 1647–56.

Prashanth, L. K., A. B. Taly, S. Sinha, G. R. Arunodaya, and H. S. Swamy. 2004. "Wilson's disease: diagnostic errors and clinical implications." *J Neurol Neurosurg Psychiatry* 75 (6): 907–9.

Travis, John. 1993. "Copper Clues Clarify Metabolic Puzzle." *Science* 262:333.

Chapter 2: Just One Bad Apple . . .

The Long Stretch Gene

Andrews, L. B. 2001. "A conceptual framework for genetic policy: comparing the medical, public health, and fundamental rights models." *Wash Univ Law Q* 79 (152): 221–85.

Houlston, R. S., and P. Parry. 1990. "Marfan syndrome." *J Med Genet* 27 (12): 791–92.

Hutchinson, S., A. Furger, D. Halliday,D. P. Judge, A. Jefferson, H. G. Dietz, H. Firth, and P. A. Handford. 2003. "Allelic variation in normal human FBN1 expression in a family with Marfan syndrome: a potential modifier of phenotype?" *Hum Mol Genet* 12 (18): 2269–76.

Lee, B, M. Godfrey, E. Vitale, H. Hori, M. G. Mattei, M. Sarfarazi, P. Tsipouras, F. Ramirez, and D. W. Hollister. 1991. "Linkage of Marfan syndrome and a phenotypically related disorder to two different fibrillin genes." *Nature* 352 (6333): 330–34.

McKusick, V. A. 1991. "The defect in Marfan syndrome." *Nature* 352 (6333): 279–81.

Magenis, R. E., C. L. Maslen, L. Smith, L. Allen, and L. Y. Sakai. 1991. "Localization of the fibrillin (FBN) gene to chromosome 15, band q21.1." *Genomics* 11 (2): 346–51.

Maslen, C. L., G. M. Corson, B. K. Maddox, R. W. Glanville, and L. Y. Sakai. 1991. "Partial sequence of a candidate gene for the Marfan syndrome." *Nature* 352 (6333): 334–37.

Pyeritz, R. E., and V. A. McKusick. 1979. Review of "The Marfan syndrome: diagnosis and management." *N Engl J Med* 300 (14): 772–77.

Ready. T. 1999. "Access to Presidential DNA denied." *Nature Medicine* 5: 859.

Reilly, Philip. 2000. *Abraham Lincoln's DNA and Other Adventures in Genetics.* Cold Spring Harbor, New York: Cold Spring Harbor Laboratory Press.

Sakai, L. Y., D. R. Keene, and E. Engvall. 1986. "Fibrillin, a new 350-kD glycoprotein, is a component of extracellular microfibrils." *J Cell Biol* 103 (6, Pt 1): 2499–509.

Schollin, J., B. Bjarke, and K. H. Gustavson. 1988. "Probable homozygotic form of the Marfan syndrome in a newborn child." *Acta Paediatr Scand* 77 (3): 452–56.

Schwartz, H. 1964. "Abraham Lincoln and the Marfan Syndrome." *JAMA* 187: 473–79.

Vecsey, George. 1988. "Remembering Flo Hyman." *New York Times,* Feb. 5.

The Dracula Gene

Arnold, W. N. 1996. "King George III's urine and indigo blue." *Lancet* 347 (9018): 1811–13.

Childs, S., B. M. Weinstein, M. A. Mohideen, S. Donohue, H. Bonkovsky, and M. C. Fishman. 2000. "Zebrafish dracula encodes ferrochelatase and its mutation provides a model for erythropoietic protoporphyria." *Curr Biol* 10 (16): 1001–4.

Cox ,T. M., N. Jack, S. Lofthouse, J. Watling, J. Haines, and M. J. Warren. 2005. "King George III and porphyria: an elemental hypothesis and investigation." *Lancet* 366 (9482): 332–35.

Warren, M. J., M. Jay, D. M. Hunt, G. H. Elder, and J. C. Rohl. 1996. "The maddening business of King George III and porphyria." *Trends Biochem Sci* 21 (6): 229–34.

The Expandable Gene

Andrew, S. E., Y. P. Goldberg, B. Kremer, H. Telenius, J. Theilmann, S. Adam, E. Starr F. Squitieri, B. Lin, M. A. Kalchman et al. 1993. "The relationship between trinucleotide (CAG) repeat length and clinical features of Huntington's disease." *Nat Genet* 4 (4): 398–403.

Cattaneo, E., D. Rigamonti, and C. Zuccato. 2000. "The enigma of Huntington's disease." *Sci Am* 287 (6): 92–97.

Duyao, M., C. Ambrose, R. Myers, A. Novelletto, F. Persichetti, M. Frontali, S. Folstein, C. Ross, M. Franz, M. Abbott et al. 1993. "Trinucleotide repeat length instability and age of onset in Huntington's disease." *Nat Genet* 4 (4): 387–92.

Huntington, G. 1872. "On Chorea." *Medical and Surgical Reporter: a Weekly Journal,* April 13, 317–21.

Li, S. H., and X. J. Li. 2004. "Huntingtin and its role in neuronal degeneration." *Neuroscientist* 10 (5): 467–75.

Snell, R. G., J. C. MacMillan, J. P. Cheadle, I. Fenton, L. P. Lazarou, P. Davies, M. E. MacDonald, J. F. Gusella, P. S. Harper, and D. J. Shaw. 1993. "Relationship between trinucleotide repeat expansion and phenotypic variation in Huntington's disease." *Nat Genet* 4 (4): 393–97.

Wexler, N. S. 2004. The U.S. Venezuela Collaborative Research Project. "Venezuelan kindreds reveal that genetic and environmental factors modulate Huntington's disease age of onset." *Proc Natl Acad Sci USA* 101 (10): 3498–503.

Chapter 3: You Can Blame It on Mom

The Gene That Launched a Revolution

Biggs, R., A. S. Douglas, R. G. MacFarlane, J. V. Dacie, W. R. Pitney, C. Merskey, and J. R. O'Brien. 1952. "Christmas disease: a condition previously mistaken for haemophilia." *Br Med J* 2 (4799): 1378–82.

Encyclopædia Britannica Online, s.v. "Nicholas II." 2005. Encyclopædia Britannica Premium Service. http://www.britannica.com/eb/article?tocId=9055725.

Haemophilia. 1997. "The history of haemophilia." Suppl. 1 of vol. 3.

Ingram, G. I.C. 1976. "The history of haemophilia." *J Clinical Pathol* 29:469–79.

MacFarlane, R. G. 1964. "An enzyme cascade in the blood clotting mechanism and its function as a biochemical amplifier." *Nature* 202:498–99.

Mannucci, P. M., and E. G. D. Tuddenham. 2001. "The hemophilias—from royal genes to gene therapy." *New Eng J Med* 344:1773–79.

McKusick, V. A. 1965. "The royal hemophilia." *Sci Am* 213 (2): 88–95.

Potts, D. M., and W. T. W. Potts. 1995. *Queen Victoria's gene: haemophilia and the Royal family*. Gloucester: Alan Sutton Publishing.

Rosner. F. 1969. "Hemophilia in the Talmud and Rabbinic writings." *Ann Intern Med* 70:833.

The Fragile X

Eberhart, D. E., H. E. Malter, Y. Feng, and S. T. Warren. 1996. "The fragile X mental retardation protein is a ribonucleoprotein containing both nuclear localization and nuclear export signals." *Hum Mol Genet* 5 (8): 1083–91.

Handa. V., T. Saha, and K. Usdin. 2003. "The fragile X syndrome repeats form RNA hairpins that do not activate the interferon-inducible protein kinase, PKR, but are cut by Dicer." *Nucleic Acids Res* 31 (21): 6243–48.

Jin, P., R. S. Alisch, and S. T. Warren. 2004. "RNA and microRNAs in fragile X mental retardation." *Nat Cell Biol* 6 (11): 1048–53.

Leehey, M. A., R. P. Munhoz, A. E. Lang, J. A. Brunberg, J. Grigsby, C. Greco, S. Jacquemont, F. Tassone, A. M. Lozano, P. J. Hagerman, and R. J. Hagerman. 2003. "The fragile X premutation presenting as essential tremor." *Arch Neurol* 60 (1): 117–21.

Lubs, H. A. 1969. "A marker X chromosome." *Am J Hum Genet* 21 (3): 231–44.
Sutherland, G. R. 1977. "Fragile sites on human chromosomes: demonstration of their dependence on the type of tissue culture medium." *Science* 197 (4300): 265–6.
Fu, Y. H., D. P. Kuhl, A. Pizzuti, M. Pieretti, J. S. Sutcliffe, S. Richards, A. J. Verkerk, J. J. Holden, R. G. Fenwick Jr, S. T. Warren et al. 1991. "Variation of the CGG repeat at the fragile X site results in genetic instability: resolution of the Sherman paradox." *Cell* 67 (6): 1047–58.

The Werewolf Gene

Figuera, L. E., M. Pandolfo, P. W. Dunne, J. M. Cantu, and P. I. Patel. 1995. "Mapping of the congenital generalized hypertrichosis locus to chromosome Xq24-q27.1." *Nat Genet* 10 (2): 202–7.
Garcia-Cruz, D., L. E. Figuera, and J. M. Cantu. 2002. "Inherited hypertrichoses." *Clin Genet* 61 (5): 321–29.
Hall, B. K. 1995. "Atavisms and atavistic mutations." *Nat Genet* 10 (2): 126–27.

The Cue Ball Gene

Hillmer, A. M., Hanneken, Ritzmann, Becker, Freudenberg, Brockschmidt, Flaquer et al. 2005. "Genetic variation in the human androgen receptor gene is the major determinant of common early-onset androgenetic alopecia." *Am J Hum Genet* 77 (1): 140–48

Chapter 4: Leaving an Imprint

Fraga, M. F., E. Ballestar, M. F. Paz, S. Ropero, F. Setien, M. L. Ballestar, D. Heine-Suner, J. C. Cigudosa, M. Urioste, J. Benitez, M. Boix-Chornet et al. 2005. "Epigenetic differences arise during the lifetime of monozygotic twins." *Proc Natl Acad Sci USA* 102 (30): 10604–9.

Whither Mom or Dad Gene

Albrecht, U., J. S. Sutcliffe, B. M. Cattanach, C. V. Beechey, D.Armstrong, G. Eichele, and A. L. Beaudet. 1997. "Imprinted expression of the murine Angelman syndrome gene, Ube3a, in hippocampal and Purkinje neurons." *Nat Gene* 17 (1): 75–78.
Clayton-Smith, J., and L. Laan. 2003. "Angelman syndrome: a review of the clinical and genetic aspects." *J Med Genet* 40 (2): 87–95.
Driscoll, D. J., M. F. Waters, C. A. Williams, R. T. Zori, C. C. Glenn, K. M. Avidano, and R. D. Nicholls.1992. "A DNA methylation imprint, determined by the sex of the parent, distinguishes the Angelman and Prader-Willi syndromes." *Genomics* 13 (4): 917–24.
Ferguson-Smith, A. C., and M. A. Surani. 2001. "Imprinting and the epigenetic asymmetry between parental genomes." *Science* 293 (5532): 1086–89.

Glenn, C. C., D. J. Driscoll, T. P. Yang, and R. D. Nicholls. 1997. "Genomic imprinting: potential function and mechanisms revealed by the Prader-Willi and Angelman syndromes." *Mol Hum Reprod* 3 (4): 321–32.

Kishino, T., M. Lalande, and J. Wagstaff. 1997. "UBE3A/E6-AP mutations cause Angelman syndrome." *Nat Genet* 15 (1): 70–73.

Nicholls, R. D., G. S. Pai, W. Gottlieb, and E. S. Cantu. 1992. "Paternal uniparental disomy of chromosome 15 in a child with Angelman syndrome." *Ann Neurol* 32 (4): 512–18.

Reik, W., A. Collick, M. L. Norris, S. C. Barton, and M. A. Surani. 1987. "Genomic imprinting determines methylation of parental alleles in transgenic mice." *Nature* 328 (6127): 248–51.

Reik, W., W. Dean, and J. Walter. 2001. "Epigenetic reprogramming in mammalian development." *Science* 293 (5532): 1089–93.

Reik, W., M. Constancia, A. Fowden, N. Anderson, W. Dean, A. Ferguson-Smith, B.

Travis, John. 1999. "Battle of the Sexes: Mouse studies shed light on whether maternal and paternal genes wage war." *Science News*. May 15.

Tycko, and C. Sibley. 2003. "Regulation of supply and demand for maternal nutrients in mammals by imprinted genes." *J Physiol* 547 (Pt 1): 35–44.

Vrana, P. B., X. J. Guan, R. S. Ingram, and S. M. Tilghman. 1998. "Genomic imprinting is disrupted in interspecific Peromyscus hybrids." *Nat Genet* 20 (4): 362–65.

The Calico Cat Gene

Brown, C. J., A. Ballabio, J. L. Rupert, R. G. Lafreniere, M. Grompe, R. Tonlorenzi, and H. F. Willard. 1991. "A gene from the region of the human X inactivation centre is expressed exclusively from the inactive X chromosome." *Nature* 349: 38–44.

Carrel, L., and H. F. Willard. 1999. "Heterogeneous gene expression from the inactive X chromosome: an X-linked gene that escapes X inactivation in some human cell lines but is inactivated in others." *Proc Natl Acad Sci USA* 96 (13): 7364–69.

———. 2005. "X-inactivation profile reveals extensive variability in X-linked gene expression in females." *Nature* 434 (7031): 400–404.

Ganesan, S.D. P. Silver, R. A. Greenberg, D. Avni, R. Drapkin, A, Miron, S. C. Mok et al. 2002. "BRCA1 supports XIST RNA concentration on the inactive X chromosome." *Cell* 111: 393–405.

Gartler, S. M., and A. D. Riggs. 1983. "Mammalian X-chromosome inactivation." *Annu Rev Genet* 17: 155–90.

Lyon, M. F. 1996. "X chromosome inactivation: pinpointing the centre." *Nature* 379: 116–17.

———. 1998. "X-chromosome inactivation: a repeat hypothesis." *Cytogenet Cell Genet* 80: 133–37.

———. 2005. "No longer 'all-or-none.'" *Eur J Hum Genet* 13 (7): 796–97.

Penny, G. D., G. F. Kay, S. A. Sheardown, S. Rastan, and N. Brockdorff. 1974. "Requirement for Xist in X chromosome inactivation." *Nature* 379: 131–37.

Therman, E., G. E. Sarto, and K. Patau1974. "Center for Barr body condensation of the proximal part of the human Xq: a hypothesis." *Chromosoma* 44: 361–66.

Travis, John. 2000. "Silence of the Xs: Does junk DNA help women muffle one X chromosome?" *Science News*. August 5.

When a Gene Won't Silence

Amir, R. E., I. B. Van den Veyver, M. Wan, C. Q. Tran, U. Francke, and H. Y. Zoghbi. 1999. "Rett syndrome is caused by mutations in X-linked MECP2, encoding methyl-CpG-binding protein 2." *Nature Genet* 23: 185–88.

Clayton-Smith, J., P. Watson, S. Ramsden,and G. C. M. Black. 2000. "Somatic mutation in MECP2 as a non-fatal neurodevelopmental disorder in males." *Lancet* 356: 830–32.

De Bona, C., M. Zappella, G. Hayek, I. Meloni, F. Vitelli, M. Bruttini, R. Cusano, P. Loffredo,I. Longo,and A. Renieri. 2000. "Preserved speech variant is allelic of classic Rett syndrome." *Europ J Hum Genet* 8:325–330.

Hagberg, B., J. Aicardi, K. Dias, and O. Ramos. 1983. "A progressive syndrome of autism, dementia, ataxia, and loss of purposeful hand use in girls: Rett's syndrome: report of 35 cases." *Ann Neurol* 14: 471–79.

Neul, J. L., and H. Y. Zoghbi. 2004. "Rett syndrome: a prototypical neurodevelopmental disorder." *Neuroscientist* 10 (2): 118–28.

Rett, A. 1986. "Rett syndrome: history and general overview." *Am J Med Genet* 1: 21–25.

Schwartzman, J. S., M. Zatz, L. R. Vasquez, R. R. Gomes, C. P. Koiffmann, C. Fridman, and P. G. Otto. 1999. "Rett syndrome in a boy with a 47, XXY karyotype." Letter. *Am J Hum Genet* 64: 1781–85.

Shahbazian, M. D., J. I. Young, L. A. Yuva-Paylor, C. M. Spencer, B. A. Antalffy, J. L. Noebels, D. L. Armstrong, R. Paylor, and H. Y. Zoghbi. 2002. "Mice with truncated MeCP2 recapitulate many Rett syndrome features and display hyperacetylation of histone H3." *Neuron* 35: 243–54.

Young J. I., E. P. Hong, J. C. Castle, J. Crespo-Barreto, A. B. Bowman, M. F. Rose, D. Kang et al. 2005. "Regulation of RNA splicing by the methylation-dependent transcriptional repressor methyl-CpG binding protein 2." *Proc Natl Acad Sci USA*. Oct. 26: 10.1073/pnas.0507856102.

Chapter 5: Just a Little Piece of the Puzzle

Speaking with a "Forked Tongue"

Chomsky, N. 1959. "Review of Verbal Behavior." *Language* 35:26–58.

Fisher, S. E., F. Vargha-Khadem, K. E. Watkins, A. P. Monaco, and M. E. Pembrey. 1998. "Localisation of a gene implicated in a severe speech and

language disorder." *Nat Genet* 18 (2): 168–70. Erratum in: 1998. *Nat Genet* 18 (3): 298.
Haesler, S., K. Wada, A. Nshdejan, E. E. Morrisey, T. Lints, E. D. Jarvis, and C. Scharff. 2004. "FoxP2 expression in avian vocal learners and non-learners." *J Neurosci* 24 (13): 3164–75.
Lai, C. S., S. E. Fisher, J. A. Hurst, F. Vargha-Khadem, and A. P. Monaco. 2001. "A forkhead-domain gene is mutated in a severe speech and language disorder." *Nature* 413 (6855): 519–23.
Pinker, S. 2001. "Talk of genetics and vice versa." *Nature* 413 (6855): 465–66.
Shu, W., J. Y. Cho, Y. Jiang, M. Zhang, D. Weisz, G. A. Elder, J. Schmeidler et al. 2005. "Altered ultrasonic vocalization in mice with a disruption in the Foxp2 gene." *Proc Natl Acad Sci USA* 102 (27): 9643–48.

The Cheeseburger Gene

Bouchard, C., L. Perusse, Y. C. Chagnon, C. Warden, and D. Ricquier. 1997. "Linkage between markers in the vicinity of the uncoupling protein 2 gene and resting metabolic rate in humans." *Hum Mol Genet* 6 (11): 1887–89.
Dulloo, A. G., J. Seydoux, and J. Jacquet. 2004. "Adaptive thermogenesis and uncoupling proteins: a reappraisal of their roles in fat metabolism and energy balance." *Physiol Behav* 83 (4): 587–602.
Fleury, C., M. Neverova, S. Collins, S. Raimbault, O. Champigny, C. Levi-Meyrueis, F. Bouillaud, M. F. Seldin, R. S. Surwit, D. Ricquier, and C. H. Warden. 1997. "Uncoupling protein-2: a novel gene linked to obesity and hyperinsulinemia." *Nat Genet* 15 (3): 269–72.
Harper, M. E., and M. F. Gerrits. 2004. "Mitochondrial uncoupling proteins as potential targets for pharmacological agents." *Curr Opin Pharmacol* 4 (6): 603–7.
Kojima, M., H. Hosoda, Y. Date, M. Nakazato, H. Matsuo, and K. Kangawa. 1999. "Ghrelin is a growth-hormone-releasing acylated peptide from stomach." *Nature* 402 (6762): 656–60.
Nicholls, D. G. 1977. "The effective proton conductance of the inner membrane of mitochondria from brown adipose tissue. Dependency on proton electrochemical potential gradient." *Eur J Biochem* 77 (2): 349–56.
Loos, R. J., and T. Rankinen. 2005. "Gene-diet interactions on body weight changes." *J Am Diet Assoc* 105 (5, Suppl 1): S29–34.
Pelleymounter, M. A., M. J. Cullen, M. B. Baker, R. Hecht, D. Winters, T. Boone, and F. Collins. 1995. "Effects of the obese gene product on body weight regulation in ob/ob mice." *Science* 269 (5223): 540–43.
Ricquier, D., G. Mory, and P. Hemon.1979. "Changes induced by cold adaptation in the brown adipose tissue from several species of rodents, with special reference to the mitochondrial components." *Can J Biochem* 57 (11): 1262–66.
Surwit, R. S., S. Wang, A. E. Petro, D. Sanchis, S. Raimbault, D. Ricquier, and S.Collins. 1998. "Diet-induced changes in uncoupling proteins in

obesity-prone and obesity-resistant strains of mice." *Proc Natl Acad Sci USA* 95 (7): 4061–65.

Yu, X., D. R. Jacobs Jr, P. J. Schreiner, M. D. Gross, M. W. Steffes, and M. Fornage. 2005. "The uncoupling protein 2 Ala55Val polymorphism is associated with diabetes mellitus: the CARDIA study." *Clin Chem* 51 (8): 1451–56.

Zhang, C. Y., G. Baffy, P. Perret, S. Krauss, O. Peroni, D. Grujic, T. Hagenet al.2001. "Uncoupling protein-2 negatively regulates insulin secretion and is a major link between obesity, beta cell dysfunction, and type 2 diabetes." *Cell* 105 (6): 745–55.

Zhang, Y., R. Proenca, M. Maffei, M. Barone, L. Leopold, and J. M. Friedman. 1994. "Positional cloning of the mouse obese gene and its human homologue." *Nature* 372 (6505): 425–32.

The Bitter Gene, or The Battle over Broccoli

Bartoshuk, L. M., V. B. Duffy, and I. J. Miller. 1994. "PTC/PROP tasting: anatomy, psychophysics, and sex effects." *Physiol Behav* 56 (6): 1165–71. Review. Erratum in: 1995. *Physiol Behav* 58 (1): 203.

Drayna, D. 2005. "Human taste genetics." *Annu Rev Genomics Hum Genet* 6:217–35.

Fackelmann, Kathleen. 1997. "The Bitter Truth: Do some people inherit a distaste for broccoli?" *Science News*. July 12.

Fisher, R. A., E. B. Ford, and J. Huxley. 1939. "Taste-testing the anthropoid apes." *Nature* 144:750.

Fox, A. L. 1932. "The relationship between chemical constitution and taste." *Proc Natl Acad Sci USA* 18:115–20.

Kim, U. K., E. Jorgenson, H. Coon, M. Leppert, N. Risch, and D. Drayna. 2003. "Positional cloning of the human quantitative trait locus underlying taste sensitivity to phenylthiocarbamide." *Science* 299:1221–25.

Kim, U. K., S. Wooding, D. Ricci, L. B. Jorde, and D. Drayna. 2005. "Worldwide haplotype diversity and coding sequence variation at human bitter taste receptor loci." *Hum Mutat* 26 (3): 199–204.

Mennella, J. A., M. Y. Pepino, and D. R. Reed. 2005. "Genetic and environmental determinants of bitter perception and sweet preferences." *Pediatrics* 115 (2): e216–22.

Wooding, S., U. K. Kim, M. J. Bamshad, J. Larsen, L. B. Jorde, and D. Drayna. 2004. "Natural selection and molecular evolution in PTC, a bitter-taste receptor gene." *Am J Hum Genet* 74 (4): 637–46.

The Schwarzenegger Gene: From Mighty Mice to Hulking Human

McPherron, A. C., A. M. Lawler, and S. J. Lee. 1997. "Regulation of skeletal muscle mass in mice by a new TGF-beta superfamily member." *Nature* 387: 83–90.

Schuelke, M., K. R. Wagner, L. E. Stolz, C. Hubner, T. Riebel, W. Komen, T. Braun, J. F. Tobin, and S. J. Lee. 2004. "Myostatin mutation associated with gross muscle hypertrophy in a child." *N Engl J Med* 350 (26): 2682–88.

Travis, John. 1997. "Muscle-bound cattle reveal meaty mutation." *Science News*." November 22.

Westhusin, M. 1997. "From mighty mice to mighty cows." *Nat Genet* 17 (1): 4–5.

Zimmers, T. A., M. V. Davies, L. G. Koniaris, P. Haynes, A. F. Esquela, K. N. Tomkinson, A. C. McPherron, N. M. Wolfman, and S. J. Lee. 2002. "Induction of cachexia in mice by systemically administered myostatin." *Science* 296 (5572): 1486–88.

A Performance Gene

Folland, J., B. Leach, T. Little, K. Hawker, S. Myerson, H. Montgomery, and D. Jones. 2000. "Angiotensin-converting enzyme genotype affects the response of human skeletal muscle to functional overload." *Exp Physiol* 85 (5): 575–79.

Jones, A., H. E. Montgomery, and D. R. Woods. 2002. "Human performance: a role for the ACE genotype?" *Exerc Sport Sci Rev* 30 (4): 184–90.

Kritchevsky, S. B., B. J. Nicklaus, M. Visser, E. M. Simonsick, A. B. Newman, T. B. Harris, E. M. Lange et al. 2005. "Angiotensin-Converting enzyme insertion/deletion genotype, exercise, and physical decline." *JAMA* 29(6): 691–98.

Marshall, R. P., S. Webb, G. J. Bellingan, H. E. Montgomery, B. Chaudhari, R. J. McAnulty, S. E. Humphries, M. R. Hill, and G. J. Laurent. 2002. "Angiotensin converting enzyme insertion/deletion polymorphism is associated with susceptibility and outcome in acute respiratory distress syndrome." *Am J Respir Crit Care Med* 166 (5): 646–50.

Tsianos G, K.I, Eleftheriou, E. Hawe, L. Woolrich, M. Watt, I. Watt, A. Peacock, H. Montgomery, and S. Grant. 2005. "Performance at altitude and angiotensin I-converting enzyme genotype." *Eur J Appl Physiol* 93 (5–6): 630–33.

An Aging Gene

Goto, M., M. Rubenstein, J. Weber, K. Woods, and D. Drayna. 1992. "Genetic linkage of Werner's syndrome to five markers on chromosome 8." *Nature* 355 (6362): 735–38.

Kipling, D., T. Davis, E. L. Ostler, and R. G. Faragher. 2004. "What can progeroid syndromes tell us about human aging?" *Science* 305 (5689): 1426–31.

Monnat, R. J. Jr., and Y. Saintigny. 2004. "Werner syndrome protein—unwinding function to explain disease." *Sci Aging Knowledge Environ* 13:3.

Rodriguez-Lopez, A. M., D. A. Jackson, F. Iborra, and L. S. Cox. 2002. "Asymmetry of DNA replication fork progression in Werner's syndrome." *Aging Cell* 1 (1): 30–39.
Wadyka, Sally. 2005. "Rapunzels of a Certain Age." *New York Times,* May 26.
Yamamoto, K., A. Imakiire, N. Miyagawa, and T. Kasahara. 2003. "A report of two cases of Werner's syndrome and review of the literature." *J Orthop Surg* (Hong Kong) 11 (2): 224–33.
Yu, C. E., J. Oshima, Y. H. Fu, E. M. Wijsman, F. Hisama, R. Alisch, S. Matthews et al. 1996. "Positional cloning of the Werner's syndrome gene." *Science* 272 (5259): 258–62.

Chapter 6: In the Beginning

The Cut-and-Paste Genes

Agrawal, A., Q. M. Eastman, and D. G. Schatz. 1998. "Transposition mediated by RAG1 and RAG2 and its implications for the evolution of the immune system." *Nature* 394 (6695): 744–51.
Jones, J. M., and M. Gellert. 2003. "The taming of a transposon: V(D)J recombination and the immune system." *Immunol Rev* 200:233–48.
Livak, F., and D. G. Schatz. 1998. "Alternative splicing of rearranged T cell receptor delta sequences to the constant region of the alpha locus." *Proc Natl Acad Sci USA* 95 (10): 5694–99.
Roth, D. B., and N. L. Craig. 1998. "VDJ recombination: a transposase goes to work." *Cell* 94 (4): 411–14.
Travis, John. 1998. "The Accidental Immune System: Long ago, a wandering piece of DNA—perhaps from a microbe—created a key strategy." *Science News*. November 7.
Yu, W., Z. Misulovin, H. Suh, R. R. Hardy, M. Jankovic, N. Yannoutsos, and M. C. Nussenzweig. 1999. "Coordinate regulation of RAG1 and RAG2 by cell type-specific DNA elements 5' of RAG2." *Science* 285 (5430): 1080–84.

Jomon Genes

Anzai, T., T. K. Naruse, K. Tokunaga, T. Honma, H. Baba, T. Akazawa, and H. Inoko. 1999. "HLA genotyping of 5,000- and 6,000-year-old ancient bones in Japan." *Tissue Antigens* 54 (1): 53–58.
Bing, S., C. Xiao, R. Deka, M. T. Seielstad, D. Kangwanpong, J. Xiao, D. Lu et al. 2000. "Y chromosome haplotypes reveal prehistorical migrations to the Himalayas." *Hum Genet* 107: 582–90.
Diamond, Jared. 1998. "The Japanese Roots." *Discover Magazine,* June.
Encyclopædia Britannica Online, s.v. "Jomon culture." 2005. Encyclopædia Britannica Premium Service, 15 Jan. 2005. http://www.britannica.com/eb/article?tocId=9043915.
———. s.v. "Kammu." Encyclopædia Britannica Premium Service. Jan. 15, 2005. http://www.britannica.com/eb/article-9044494.

———. s.v. "The Peopling of Japan." Encyclopædia Britannica Premium Service. Jan. 15, 2005. http://www.britannica.com/eb/article?tocId=9114375.

Hammer, M. F., and S. Horai. 1995. "Y chromosomal DNA variation and the peopling of Japan." *Am J Hum Genet* 56 (4): 951–62.

Jin, H. J., K. D. Kwak, M. F. Hammer, Y. Nakahori, T. Shinka, J. W. Lee, F. Jin, X. Jia, C. Tyler-Smith, and W. Kim. 2003. *Hum Genet* 114 (1): 27–35.

Matsubara, Hiroshi. 2001. "Emperor's Remark Pours Fuel on Ethnic Hot Potato." *Japan Times*, March 12.

Niigata Prefectural Museum of History. http://www.nbz.or.jp/jp/index.html.

Tanaka, M., V. M. Cabrera, A. M. González, J. M. Larruga, T. Takeyasu, N. Fuku, L.-J. Guo et al. 2004. "Mitochondrial genome variation in eastern Asia and the peopling of Japan." *Genome Res* 10A (Oct. 14) :1832–50.

Travis, John. 1997. "Jomon Genes: Using DNA, researchers probe the genetic origins of modern Japanese." *Science News,* Feb. 15.

Weiner, Eric. 2001. "Life as an 'Outside Person.'" *National Public Radio,* Sept. 4.

Survivors' Benefit

Dalgleish, A. G., P. C. Beverley, P. R. Clapham, D. H. Crawford, M. F. Greaves, and R. A. Weiss. 1984. "The CD4 (T4) antigen is an essential component of the receptor for the AIDS retrovirus." *Nature* 312 (5996): 763–37.

Dean, M., Carrington, Winkler, Huttley, Smith, Allikmets, Goedert et al. 1996. "Genetic restriction of HIV-1 infection and progression to AIDS by a deletion allele of the CKR5 structural gene." Hemophilia Growth and Development Study, Multicenter AIDS Cohort Study, Multicenter Hemophilia Cohort Study, San Francisco City Cohort, ALIVE Study. *Science* 273 (5283): 1856–62. Erratum in: 1996. *Science* 274 (5290): 1069.

Duncan, C. J., and S. Scott. 2005. "What caused the Black Death?" *Postgrad Med J* 81 (955): 315–20.

Duncan, S. R., S. Scott, and C. J. Duncan. 2005. "Reappraisal of the historical selective pressures for the CCR5-Delta32 mutation." *J Med Genet* 42 (3): 205–8.

Galvani, A. P., and M. Slatkin. 2003. "Evaluating plague and smallpox as historical selective pressures for the CCR5-Delta 32 HIV-resistance allele." *Proc Natl Acad Sci USA* 100 (25): 15276–79.

Lalani A. S., J. Masters, W. Zeng, J. Barrett, R. Pannu, H. Everett, C. W. Arendt, and G. McFadden. 1999. "Use of chemokine receptors by poxviruses." *Science* 286 (5446): 1968–71.

Libert, F., P. Cochaux, G. Beckman, M. Samson, M. Aksenova, A. Cao, A. Czeizel et al. 1998. "The deltaccr5 mutation conferring protection against HIV-1 in Caucasian populations has a single and recent origin in Northeastern Europe." *Hum Mol Genet* 7 (3): 399–406.

Liu, R., W. A. Paxton, S. Choe, D. Ceradini, S. R. Martin, R. Horuk, M. E. MacDonald, H. Stuhlmann, R. A. Koup, and N. R. Landau. 1996. "Homozygous defect in HIV-1 coreceptor accounts for resistance of some multiply-exposed individuals to HIV-1 infection." *Cell* 86 (3): 367–77.

Mecsas, J., G. Franklin, W. A. Kuziel, R. R. Brubaker, S. Falkow, and D. E. Mosier. 2004. "Evolutionary genetics: CCR5 mutation and plague protection." *Nature* 427 (6975): 606.

The Pregnancy Genes

Blond, J.-L., F. Beseme, L. Duret, O. Bouton, F. Bedin, H. Perron, B. Mandrand, and F. Mallet. 1999. "Molecular characterization and placental expression of HERV-W, a new human endogenous retrovirus family." *J. Virol* 73:1175–85.

Blond, J.-L., D. Lavillette, V. Cheynet, O. Bouton, G. Oriol, S. Chapel-Fernandes, B. Mandrand, F. Mallet, and F.-L. Cosset. 2000. "An envelope glycoprotein of the human endogenous retrovirus HERV-W is expressed in the human placenta and fuses cells expressing the type D mammalian retrovirus receptor." *J. Virol* 74:3321–29.

Fox, Douglas. 1999. "Why we don't lay eggs." *New Scientist*. June 12.

Frendo, J. L., D. Olivier, V. Cheynet, J.-L. Blond, O. Bouton, M. Vidaud, M. Rabreau, D. Evain-Brion, and F. Mallet. 2003. "Direct involvement of HERV-W Env glycoprotein in human trophoblast cell fusion and differentiation." *Mol Cell Biol* 23 (10): 3566–74.

Knerr, I., E. Beinder, and W. Rascher. 2002. "Syncytin, a novel human endogenous retroviral gene in human placenta: evidence for its dysregulation in preeclampsia and HELLP syndrome." *Am J Obstet Gynecol* 186 (2): 210–13.

Lee, X., J. C. Keith Jr, N. Stumm, I. Moutsatsos, J. M. McCoy, C. P. Crum, D. Genest et al. 2001. "Downregulation of placental syncytin expression and abnormal protein localization in pre-eclampsia." *Placenta* 22 (10): 808–12.

Lin, L., B. Xu, and N. S. Rote. 1999. "Expression of endogenous retrovirus ERV-3 induces differentiation in BeWo, a choriocarcinoma model of human placental trophoblast." *Placenta* 20 (1): 109–18.

Mi, S. X. Lee, X. Li, G. M. Veldman, H. Finnerty, L. Racie, E. LaVallie et al. 2000. "Syncytin is a captive retroviral envelope protein involved in human placental morphogenesis." *Nature* 403:785–89.

Muir, A., A. Lever, and A. Moffett. 2004. "Expression and functions of human endogenous retroviruses in the placenta: an update." *Placenta* 25 (Suppl A): S16–25.

O'Connell, C., S. O'Brien, W. G. Nash, and M. Cohen. 1984. "ERV3, a full-length human endogenous provirus: chromosomal localization and evolutionary relationships." *Virology* 138:225–35.

Rote, N. S., S. Chakrabarti, and B. P. Stetzer. 2004. "The role of human endogenous retroviruses in trophoblast differentiation and placental development." *Placenta* 25 (8–9): 673–83.

Travis, John. 2000. "Placental Puzzle: Do captured viral genes make human pregnancies possible?" *Science News*. May 13.

The "Got Milk?" Gene

Boll, W., P. Wagner, and N. Mantei. 1991. "Structure of the chromosomal gene and cDNAs coding for lactase-phlorizin hydrolase in humans with adult-type hypolactasia or persistence of lactase." *Am J Hum Genet* 48 (5): 889–902.

Bersaglieri, T., P. C. Sabeti, N. Patterson, T. Vanderploeg, S. F. Schaffner, J. A. Drake, M. Rhodes, D. E. Reich, and J. N. Hirschhorn. 2004. "Genetic signatures of strong recent positive selection at the lactase gene." *Am J Hum Genet* 74 (6): 1111–20.

Enattah, N. S., T. Sahi, E. Savilahti, J. D. Terwilliger, L. Peltonen, and I. Jarvela. 2002. "Identification of a variant associated with adult-type hypolactasia." *Nat Genet* 30 (2): 233–37.

Mulcare, C. A., M. E. Weale, A. L. Jones, B. Connell, D. Zeitlyn, A. Tarekegn, D. M. Swallow, N. Bradman, and M. G. Thomas. 2004. "The T allele of a single-nucleotide polymorphism 13.9 kb upstream of the lactase gene (LCT) (C-13.9kbT) does not predict or cause the lactase-persistence phenotype in Africans." *Am J Hum Genet* 74 (6): 1102–10.

Troelsen, J. T. 2005. "Adult-type hypolactasia and regulation of lactase expression." *Biochim Biophys Acta* 1723 (1–3): 19–32.

Innate Sensing Genes

Beutler, B., K. Hoebe, P. Georgel, K. Tabeta, and X. Du. 2005. "Genetic analysis of innate immunity: identification and function of the TIR adapter proteins." *Adv Exp Med Biol* 560:29–39.

Blander, J. M., and R. Medzhitov. 2004. "Regulation of phagosome maturation by signals from toll-like receptors." *Science* 304 (5673): 1014–18.

Ito, T., R. Amakawa, T. Kaisho, H. Hemmi, K. Tajima, K. Uehira, Y. Ozaki, H. Tomizawa, S. Akira, and S. Fukuhara. 2002. "Interferon-alpha and interleukin-12 are induced differentially by Toll-like receptor 7 ligands in human blood dendritic cell subsets." *J Exp Med* 195 (11): 1507–12.

Lederberg, Joshua, Robert E. Shope, and Stanley C. Oaks, Jr. 1992. *Emerging Infections: Microbial Threats to Health in the United States.* Washington, DC: National Academy Press.

Medzhihtov, R., and C. Janeway Jr. 1997. "Innate immunity: the virtues of a nonclonal system of recognition." *Cell* 91:295–98.

Mellman, I., and R. M. Steinman. 2001. "Dendritic cells: specialized and regulated antigen processing machines." *Cell* 106:255–58.

Pasare, C., and R. Medzhitov R. 2005. "Toll-like receptors: linking innate and adaptive immunity." *Adv Exp Med Biol* 560:11–18.

Poltorak, A., X. He, I. Smirnova, M. Y. Liu, C. Van Huffel, X. Du, D. Birdwell et al. 1998. "Defective LPS signaling in C3H/HeJ and C57BL/10ScCr mice: mutations in Tlr4 gene." *Science* 282 (5396): 2085–88.

Ries, L. A. G., M. P. Eisner, C. L. Kosary, B. F. Hankey, B. A. Miller, L. Clegg, A. Mariotto, E. J. Feuer, and B. K. Edwards eds. *SEER Cancer Statistics*

Review, 1975–2002. Bethesda, MD: National Cancer Institute, http://seer.cancer.gov/csr/1975_2002/, based on Nov. 2004 SEER data submission, posted to the SEER web site 2005.

Travis, John. 2001. "Immunity's Eyes: Biologists reveal the proteins that first see dangerous microbes." *Science News*. September 8.

The Sidedness Genes

Ainsworth, Claire. 2000. "Left Right and Wrong." *New Scientist*. June 17.

Brody, S. L. 2004. "Genetic regulation of cilia assembly and the relationship to human disease." *Am J Respir Cell Mol Biol* 30 (4): 435–37.

Brueckner, M. 2001. "Cilia propel the embryo in the right direction." *Am J Med Genet* 101 (4): 339–44.

Hirokawa, N. 2000. "Determination of Left-Right Asymmetry: Role of Cilia and KIF3 Motor Proteins." *News Physiol Sci* 15:56.

Hornstein, E., and C. J. Tabin. 2005. "Developmental biology: asymmetrical threat averted." *Nature* 435 (7039): 155–56.

Kawakami, Y., A. Raya, R. M. Raya, C. Rodriguez-Esteban, and J. C. Belmonte. 2005. "Retinoic acid signalling links left-right asymmetric patterning and bilaterally symmetric somitogenesis in the zebrafish embryo." *Nature* 435 (7039): 165–71.

Levin, M. 2004. "The embryonic origins of left-right asymmetry." *Crit Rev Oral Biol Med* 15 (4): 197–206.

Tabin, C. J., and K. J. Vogan. 2003. "A two-cilia model for vertebrate left–right axis specification." *Genes Dev* 17 (1): 1–6.

Tabin, C. J. 2005. "Do we know anything about how left–right asymmetry is first established in the vertebrate embryo?" *J Mol Histol* 36: 235–41.

Travis, John. 1999. "Twirl Those Organs into Place: Getting to the heart of how a heart knows left from right." *Science News*. August 21.

Epilogue: Snippets of Information Reveal the Whole?

Altshuler, D., and A. G. Clark. 2005. "Genetics. Harvesting medical information from the human family tree." *Science* 307 (5712): 1052–53.

Duster, T. 2005. "Medicine. Race and reification in science." *Science* 307 (5712): 1050–51.

Hinds, D. A., L. L. Stuve, G. B. Nilsen, E. Halperin, E. Eskin, D. G. Ballinger, K. A. Frazer, and D. R. Cox. 2005. "Whole-genome patterns of common DNA variation in three human populations." *Science* 307 (5712): 1072–79.

Appendix: A Genetics Primer

Siepel, A., G. Bejerano, J. S. Pedersen, A. S. Hinrichs, M. Hou, K. Rosenbloom, H. Clawson et al. 2005. "Evolutionarily conserved elements in vertebrate, insect, worm, and yeast genomes." *Genome Res* 15 (8): 1034–50.

Index